Nora Kristen-Hochrein

Interactions in thin liquid films

Nora Kristen-Hochrein

Interactions in thin liquid films

Oppositely charged polyelectrolyte/surfactant mixtures

Südwestdeutscher Verlag für Hochschulschriften

Imprint

Any brand names and product names mentioned in this book are subject to trademark, brand or patent protection and are trademarks or registered trademarks of their respective holders. The use of brand names, product names, common names, trade names, product descriptions etc. even without a particular marking in this work is in no way to be construed to mean that such names may be regarded as unrestricted in respect of trademark and brand protection legislation and could thus be used by anyone.

Publisher:
Südwestdeutscher Verlag für Hochschulschriften
is a trademark of
Dodo Books Indian Ocean Ltd., member of the OmniScriptum S.R.L Publishing group
str. A.Russo 15, of. 61, Chisinau-2068, Republic of Moldova Europe
Printed at: see last page
ISBN: 978-3-8381-2441-4

Zugl. / Approved by: Berlin, TU, Diss., 2010

Copyright © Nora Kristen-Hochrein
Copyright © 2011 Dodo Books Indian Ocean Ltd., member of the OmniScriptum S.R.L Publishing group

Contents

1 **Introduction**   3

2 **Scientific Background**   5
   2.1 Thin aqueous films 5
       2.1.1 Electrostatic double-layer forces 6
       2.1.2 Van der Waals forces 7
       2.1.3 Steric forces 7
       2.1.4 Structural forces 8
       2.1.5 Disjoining pressure isotherms 8
       2.1.6 Simulation of the disjoining pressure isotherms 9
   2.2 Polyelectrolyte/surfactant mixtures in foam films 10
       2.2.1 Pure surfactant films 11
       2.2.2 Likely charged polyelectrolyte/surfactant systems 12
       2.2.3 Oppositely charged polyelectrolyte/surfactant systems 12
       2.2.4 Mixtures of nonionic surfactant and charged polyelectrolytes . 15
       2.2.5 Stratification phenomena 15

3 **Experimental section**   19
   3.1 Materials 19
       3.1.1 Surfactants 19
       3.1.2 Polyelectrolytes 19
       3.1.3 Salts and Monomers 20
   3.2 Methods 20
       3.2.1 Thin Film Pressure Balance (TFPB) 20
       3.2.2 Surface Characterisation 23

4 **Effect of surface charge on foam film stability**   27
   4.1 Introduction 27
   4.2 Results 28
   4.3 Discussion 31
       4.3.1 Below the nominal isoelectric point. 31
       4.3.2 At the nominal isoelectric point. 32
       4.3.3 Above the isoelectric point. 32
       4.3.4 Foam film stabilities. 33
   4.4 Conclusions 33

5 **Effect of surfactant and polyelectrolyte hydrophobicity**   35
   5.1 Introduction 35
   5.2 Results 37
       5.2.1 $C_{12}TAB$/PAMPS mixtures 37
       5.2.2 $C_{12}TAB$/PSS mixtures 39
       5.2.3 $C_{14}TAB$/PSS mixtures 40

5.3	Discussion	41
	5.3.1 Influence of the surfactant	42
	5.3.2 Influence of the polyelectrolyte	44
5.4	Conclusions	45

6 Variation of the isoelectric point 47
- 6.1 Introduction 47
- 6.2 Results 48
- 6.3 Discussion 54
  - 6.3.1 Effect of the surfactant concentration 55
  - 6.3.2 Effect of a lower degree of polymer charge . . 56
- 6.4 Conclusion 59

7 Polyelectrolyte versus monomer effect 61
- 7.1 Introduction 61
- 7.2 Results 62
- 7.3 Discussion 72
  - 7.3.1 Salt effect 72
  - 7.3.2 Hydrophilic/hydrophobic balance 75
  - 7.3.3 Comparison between monomer and polymer . 80
- 7.4 Conclusion 82

8 Effect of the polyelectrolyte chain length 83
- 8.1 Introduction 83
- 8.2 Results 84
  - 8.2.1 PSS60/C_{12}TAB . . . 84
  - 8.2.2 PSS20/C_{12}TAB . . . 86
- 8.3 Discussion 89
- 8.4 Conclusion 92

9 Dynamics of polymer chains in thin films 95
- 9.1 Introduction 95
- 9.2 Fluorescence spectroscopy on foam films 96
  - 9.2.1 Additional experimental details 97
  - 9.2.2 Results and discussion 97
  - 9.2.3 Conclusion 100
- 9.3 Diffusion of polyelectrolytes in thin films 100
  - 9.3.1 Additional experimental details 101
  - 9.3.2 Results and discussion 104
  - 9.3.3 Conclusion 106

10 Conclusion and Outlook 107

1 Introduction

In the present thesis, the interactions in foam films from oppositely charged polyelectrolyte/ surfactants mixtures are studied. These mixtures in aqueous foams are used in many applications like detergency, cosmetics, fire fighting, and enhanced oil recovery.[1] In some applications, stable foam is desired, like for shampoo, where the customers like to have rich foam, while in others, it should be avoided. Washing agent, for example, should have a low foaming ability to avoid overflowing washing machines.

Foams are dispersions of air bubbles in a liquid. Depending on the air/water ratio in the dispersion, different foams are formed. In wet foam, the water content is very high and the bubbles are spherically shaped. When the foam ages, the liquid in the foam lamellas drains due to gravitation and dry foams emerge. In that case, the air bubbles no longer form spheres but polyhedrons so that they are often referred to as polyhedral foams. These polyhedrons correspond to the minimal surfaces, which are necessary to minimize the surface energy. In dry foams the volume fraction of air exceeds 75 %.

It is not possible to obtain stable foams from pure liquids. Stable foams are only formed in the presence of an appropriate surface active agent like surfactants, colloidal particles, polymeric surfactants, phospholipids, or proteins.[2] Additionally, energy is needed to create the bubble surfaces in the foam. As a consequence, foams are in an absolute sense thermodynamically unstable. However, there are systems that are stable for minutes or even days or weeks. In rapidly coalescing dispersions, the film lifetime is controlled by the drainage rate of the continuous phase, while the long-lived systems require additional time to overcome energy barriers that hold the film in a metastable thermodynamic state. These barriers arise from surface force interactions created by having two interfaces in close proximity.[3]

The physicochemical properties of foam and foam films have attracted scientific interest for a long time. The first recorded observations using soap films were reported by Hooke and Newton in the late 17th century. Already these works contain observations on black spots in soap films. The first systematic study of the various properties of soap films has been conducted by the Belgian scientist Plateau. He was the first to draw attention to that part of the foam that connects the single foam lamellas, which are now called Plateau borders. The theory was further developed by the thermodynamic descriptions of thin films by Gibbs and Marangoni. Further progress in the foam film research was achieved in the second half of the 20th century. By this time, many scientists like, for example, Derjaguin, Mysels, and Scheludko contributed the contemporary understanding of foam films and foams.

The properties of a foam film can be easily tuned by adding polyelectrolytes or salt to the system or by varying the experimental conditions like temperature, concentration, or pH etc. To understand the behaviour of a foam it is crucial to investigate the properties of the thin liquid films that separate the gas compartments of the dispersed phase. Depending on the interactions in these foam films, they have a thickness of 5-120 nm. Due to this fact, the free-standing film also corresponds to a slit pore geometry which allows the investigation of the effect of confinement on the structuring of colloidal particles, aggregates or macromolecules.

The addition of polyelectrolytes to surfactant solutions is especially interesting for many applications and has been the subject of many studies. The properties of these mixtures can be

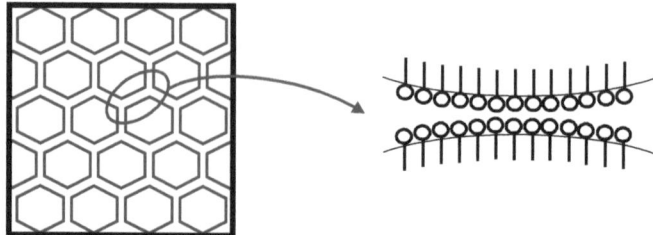

Figure 1.1: *Schematic drawing of polyhedral foam and the corresponding foam film stabilised by surfactant molecules.*

adjusted by simply varying the charge of the respective components: When the surfactant and the polyelectrolyte are nonionic, the interactions are only weak. On the other hand, when the two species are likely or oppositely charged, a strong repulsion and attraction, respectively occurs. In case of oppositely charged mixtures, polyelectrolyte and surfactant can form complexes in the bulk and at the surface, but unlike in monolayers of insoluble molecules, the adsorbed amount at the surface is not known.[4]

The main objective of this thesis is the investigation of foam films of mixtures of positively charged surfactants and negatively charged polyelectrolytes. At low and high polyelectrolyte concentrations, respectively, the foam film behaviour is well established. Pure surfactant solutions form thick common black films (CBF) due to the positive charge at the surface and the resulting electrostatic repulsion of the two opposing interfaces. Above the isoelectric point (IEP), when the polyelectrolyte concentration exceeds the surfactant concentration, a CBF is formed as well because of the strong repulsion between the polyelectrolyte chains. The question arises, how the foam films behave between the two described concentration regimes: Does a charge reversal take place at the interface? What happens at the IEP of the system, can stable films still be formed at low surface charges? And how do parameters like surfactant concentration, polyelectrolyte chain length or charge density influence the foam film characteristics?

To answer these questions, several different polyelectrolyte/surfactant mixtures are studied. In chapters 4 and 5 the influence of the polyelectrolyte and surfactant hydrophobicity is investigated, in chapter 7 and 8 the effect of polyelectrolyte chain length and of the monomer is discussed and in chapter 6 the influence of the polyelectrolyte charge density as well as the surfactant concentrations is investigated. Furthermore, in chapter 9, the dynamics of polyelectrolytes in the foam film core are studied with the help of fluorescent labels. Each chapter of the thesis is written as a complete unit and is independent of the others.

2 Scientific Background

2.1 Thin aqueous films

The stability of colloidal dispersions strongly depends on the properties of thin films of the continuous phase that separate the dispersed phase into compartments. These thin films, foam lamellas in case of a foam, are stabilised by an excess pressure with respect to the bulk liquid normal to the film surfaces, the disjoining pressure Π. It can either be repulsive ($\Pi > 0$) or attractive ($\Pi < 0$), in the latter case called conjoining pressure. The force arises from a thin transition region at the interface whose properties derivate from those of the two neighbouring bulk phases. It can be thermodynamically described by the negative derivative of the Gibbs energy by the film thickness:

$$\Pi(h) = -\left(\frac{\partial G}{\partial h}\right)_{T,P,A,n} \tag{2.1}$$

In the 1940s two groups of scientists (Derjaguin and Landau, Verwey and Overbeek) independently developed a quantitative theoretical analysis of the problem of colloidal stability. The theory is known as DLVO theory, after the initial letter of their names, and considers two types of forces: The electrostatic double-layer forces, that is always present when the surfaces contain charged groups and the van der Waals force, which operates irrespective of the chemical nature of the molecules.[5]

$$\Pi(h) = \Pi_{el} + \Pi_{vdW} \tag{2.2}$$

However, experiments have shown, that not all interactions in thin films can be explained by the DLVO theory and additional forces have to be taken into account. Since the disjoining pressure is an additive force, the various contributions of the disjoining pressure can be separated into different components:

$$\Pi(h) = \Pi_{el} + \Pi_{vdW} + \Pi_{steric} + \Pi_{struc} + ... \tag{2.3}$$

where the subscript indicate the following contributions: el = electrostatic double-layer forces, vdW = van der Waals forces, $steric$ = steric or entropic forces, and $struc$ = structural forces.

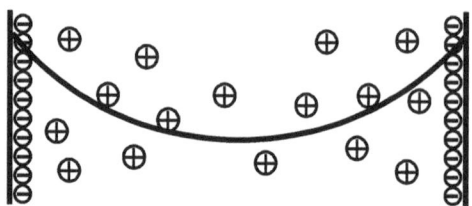

Figure 2.1: *Two charged interfaces with their diffuse double-layer; the counterion concentration is depicted by the solid line.*

2.1.1 Electrostatic double-layer forces

The first component of the disjoining pressure arises from the overlap of two electrostatic double-layers that develop at charged interfaces. When the distance between the two surfaces is in the range of the Debye length, the characteristic thickness of the diffuse double-layer, an additional force is needed to further approach the interfaces.

The Debye length can be calculated from the classical Debye-Hückel theory:

$$\frac{1}{\kappa} = \sqrt{\frac{\varepsilon_0 \varepsilon k T}{e^2 N_a \sum Z_i^2 c_i}} \qquad (2.4)$$

where ε is the dielectric constant of the medium, k is the Boltzmann constant, e the elementary charge, N_a Avogadro's number, and Z, and c, respectively, the valency and the concentration of corresponding ions. The Debye length is very sensitive to the ionic strength and decreases with increasing ion concentration.

The electrostatic component of the disjoining pressure is described by:

$$\Pi_{el} = \Pi_0 \exp(-\kappa h) \qquad (2.5)$$

To quantify the electrostatic force, the Poisson-Boltzmann equation has to be solved under certain boundary conditions. Assuming a low constant surface potential (< 50 mV) or large distances a linearised form of the Poisson-Boltzmann equation can be applied. In that case, the electrostatic double-layer force is given by:[3,6]

$$\Pi_{el} = 64 k T \rho_\infty \gamma^2 \exp(-\kappa h) \qquad (2.6)$$

where

$$\gamma = \tanh\left(\frac{ze\Psi_0}{4kT}\right) \qquad (2.7)$$

and ρ_∞ is the number density of the ions. According to Eq. 2.5, Π_0 is then given by:

$$\Pi_0 = 64 k T \rho_\infty \gamma^2 \qquad (2.8)$$

From the surface potential Ψ_0, the surface charge density can be directly calculated by using the Graham equation:[7]

2.1 Thin aqueous films

$$\sigma = \sqrt{8\varepsilon\varepsilon_0 kT}\sinh\left(\frac{e\Psi_0}{2kT}\right)\left(\sum c_i\left(2 + \exp\left(-\frac{e\Psi_0}{kT}\right)\right)\right)^{1/2} \quad (2.9)$$

In symmetric films like foam films, the two interfaces are always likely charged. This leads to a repulsive contribution of the electrostatic double-layer to the disjoining pressure and therefore to the stabilisation of the film. This effect can also be understood in terms of the osmotic pressure, that is created by the difference in ionic concentration between the two approaching surfaces and the bulk, that prevents a further approach of the interfaces.[6]

Electrostatic forces occur in foam films with non-ionic surfactants as well, where the charge can not originate from the charge of the surfactant. This leads to the conclusion, that the pure air/water interface has to carry charges as well. Experiments have shown[8,9] that the interface is slightly negatively charged due to the adsorption of OH^-, so that the pH at the interface is different from that of the bulk phase.

2.1.2 Van der Waals forces

The other important contribution to the DLVO theory, is the van der Waals force. It considers dipole-dipole interactions, interactions between dipoles and induced dipoles, and the most important contribution, the London dispersion forces between two induced dipoles. The latter describe very weak interactions present between all pairs of molecules, even between neutral species. In neutral molecules, dipoles are temporarily induced by the instantaneous position of the electrons about the nuclear protons. The instantaneous dipole generates an electric field that induces a dipole in any nearby atom.[7] This plays an important role in adhesion, adsorption, wetting, and of course in thin liquid films.

The calculation of the van der Waals component of the disjoining pressure was introduced by Hamaker and is based on the pairwise summation of the individual dispersion interactions between all molecules. It is described by the following equation:

$$\Pi_{vdW} = -\frac{A}{6\pi h^3} \quad (2.10)$$

where A is the Hamaker constant, which is characteristic for a system of two media, separated by a thin film. For symmetric films like foam films, the Hamaker constant is always positive ($A = 3.7 \times 10^{-20}$ J[5,7]), which leads to an attractive van der Waals force. In the case of a thin film entrapped between two different media, the solution is more complex and a repulsive van der Waals force can occur as well.

This force is short range (≈ 10 nm), compared to the electrostatic double layer force, due to the dependency $\Pi_{vdW} \sim h^{-3}$. It comes into account when the electrostatic barrier is overcome and can lead to the rupture of a film.

2.1.3 Steric forces

At very small distances between two surfaces, the interactions in thin films can no longer be described by the DLVO theory. Therefore, an additional force has to be introduced to the theory, the steric or entropic force. In a first attempt, this force was defined as the steric repulsion that arises from the overlap of two adsorbed layers,[10] but the origin of this force is more complex. Israelachvili and Wennerström[11] have shown, that, besides the overlapping of amphiphile head groups, a number of fluctuations of the interface contribute to this component of the disjoining pressure as well. These fluctuations include undulations, peristaltic fluctuations, and protrusion. At separations smaller than 2 nm, protrusion and headgroup overlap dominate the steric force,[3,7] while undulations have a longer range.

2.1.4 Structural forces

In addition to DLVO and steric forces, structural forces play a role in thin films. When a fluid is confined between two interfaces, a layering of the entrapped molecules occurs close to the interface. For example, micelles, colloidal particles, or polyelectrolytes have the ability to form these structures. For polyelectrolytes, this is only valid in the semi-dilute concentration regime and will be further explained in chapter 2.2. The layering is related to an oscillatory decay of the particle or molecule concentration from the interface towards the film bulk and induces a damped oscillatory disjoining pressure. This oscillatory disjoining pressure is described by an exponentially decaying cosine function:

$$\Pi_{struc} = A \exp(-\frac{h}{\lambda}) \cos(\frac{2\pi h}{d}) \qquad (2.11)$$

where A is the oscillatory amplitude, λ the decay length and d the oscillatory period.

When the two interfaces approach each other, a layer-wise expulsion of molecules from the film occurs. This is manifested by a stepwise thinning of the film, that is called stratification process. The size of the steps, Δh, is correlated to a characteristic length scale of the system, for example the effective diameter of a micelle or the mesh size of a polymer network and scales with the concentration of the structure inducing molecules. For spherical particles or micelles Δh scales with $c^{-1/3}$, while for linear polyelectrolytes, the power law has an exponent of $-1/2$.

2.1.5 Disjoining pressure isotherms

To get information about the predominant forces in a foam film, equilibrium disjoining pressure isotherms are measured, which is the disjoining pressure versus the film thickness. Such an isotherm is shown in Fig. 2.2 as a schematic representation. It is characteristic for each system and depends strongly on parameters like surfactant concentration, and additives like polyelectrolytes or salts etc. The addition of all described components of the disjoining pressure leads to a non-monotonous force. Only the parts with a negative slope are mechanically stable foam films, which divides the isotherm into two mechanically stable regions where two different types of films are formed.

Common black films (CBF) have a thickness of about 10 to 100 nm and are mainly stabilised by electrostatic forces. Very thin films with a thickness of 5 to 10 nm are called Newton black films (NBF) and are mainly stabilised by steric forces. They consist only of two surfactant layers adsorbed at the opposing film interfaces and the corresponding hydration water, while all other liquid is pressed out of the film.

A draining film can either rupture, due to attractive van der Waals forces, or a transition to a NBF can take place (cf. Fig. 2.3). In that case, thin black spots occur that spread over the whole film within several seconds. In the bright spots, the excess liquid from the NBF is transported to the liquid reservoir.

In the 'Thin Film Pressure Balance' technique that is mainly used in this thesis, only the mechanically stable parts of the disjoining pressure isotherm are accessible.

2.1 Thin aqueous films

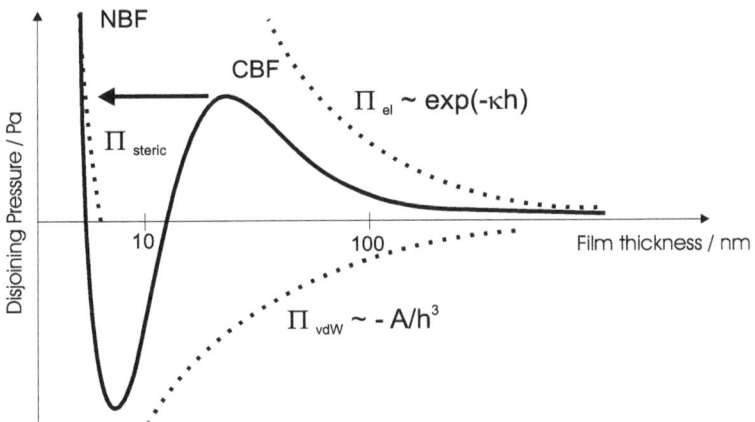

Figure 2.2: *Schematic representation of a disjoining pressure isotherm.*

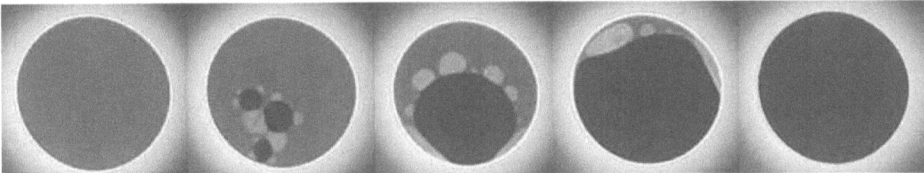

Figure 2.3: *Transition from a CBF to a NBF.*

2.1.6 Simulation of the disjoining pressure isotherms

Charged planar surfaces are characterised by a surface charge density σ and a potential Ψ_0. The liquid that is in contact with the planar surface has an impact on the two parameters due to electrolyte in the solution. The solution is characterised by the bulk concentration $c_{i,bulk}$ of the electrolyte, the ion valency z_i and the dielectric constant ε_r.[5]

The Poisson-Boltzmann equation describing the ion distribution in an electrolyte solution outside a charged interface is derived from the Poisson equation:

$$\varepsilon_0 \varepsilon_r \nabla^2 \Psi = -\rho \qquad (2.12)$$

where ∇ is the Laplace operator and ρ is the charge distribution of the free ions in the solution. According to the Boltzmann equation, the ion density can be expressed as:

$$\rho \cong e \sum z_i c_{i,bulk} \exp\left(-z_i e \Psi / kT\right) \qquad (2.13)$$

This results in the Poisson-Boltzmann equation:

$$\varepsilon_0 \varepsilon_r \nabla^2 \Psi = -e \sum_i z_i c_i \qquad (2.14)$$

with

$$c_i = c_{i,bulk} \exp(-ez_i \Psi / kT) \tag{2.15}$$

To get information about the surface potential of the foam films, the disjoining pressure isotherms are simulated with the PB program written by Per Linse.[12] In this program, the nonlinear Poisson-Boltzmann equation is solved under the assumption of constant potential. The numerical approach in the program considers two planar surfaces separated by the distance D and with their normals in z direction. Since in case of infinite equally charged planar surfaces the potential cannot change in the x and y direction because of the symmetry, only the variation in the z direction is important.[13]

$$\varepsilon_0 \varepsilon_r \frac{d\Psi^2(z)}{d^2 z} = -e \sum_i z_i c_i(z) \tag{2.16}$$

The intervening liquid is in equilibrium with a bulk electrolyte solution. Because of the symmetry of the thin film, only one half of the system needs to be considered, $0 \leq z \leq b, b = D/2$. To obtain a a unique solution, two boundary conditions are needed.[5,14] The first one follows the symmetry requirement that the field must vanish at the midplane:

$$\left. \frac{d\Psi(z)}{dz} \right|_{z=b} = 0 \tag{2.17}$$

The second boundary condition derives from the requirement of electroneutrality, *i.e.* that the total charge of the counterions between the two surfaces must be equal to the charge at the surface:

$$\left. \frac{d\Psi(z)}{dz} \right|_{z=0} = -\frac{\sigma}{\varepsilon_0 \varepsilon_r} \tag{2.18}$$

Before the simulations, the disjoining pressure isotherms are fitted with an exponential decay function of first order. From this fit, the Debye length and the corresponding ionic strength are calculated to get a good starting point for the simulation. During the simulation, the surface potential and the ionic strength are adjusted until the simulated curve coincides with the experimental data.

2.2 Polyelectrolyte/surfactant mixtures in foam films *

As already mentioned, polyelectrolyte/surfactant mixtures in foams are of great relevance in many practical applications, like personal care or cleaning. In this section, recent work on this field is reviewed to give an introduction to the subject. Firstly, the charge combination of the mixture has a great impact on the foam film. Depending on the charge of the used polyelectrolytes and surfactants, either a CBF or a NBF is formed as a final state before film rupture. Table 2.1 gives an overview about the type of foam films, that are formed at different polyelectrolyte/surfactant charge combinations. Since the main topic of this thesis is the investigation of foam films from oppositely charged polyelectrolyte/surfactant mixtures, a particular attention has been given to these systems.

For foam film studies, it is important that the concentrations of polyelectrolyte and surfac-

---

* Similar content has been published in: *Effect of polyelectrolyte/surfactant combinations on the stability of foam films*, N. Kristen and R. v. Klitzing, *Soft Matter*, **2010**, *6*, 849 - 861

2.2 Polyelectrolyte/surfactant mixtures in foam films

polymer	surfactant		
	cationic	nonionic	anionic
cationic	CBF	NBF	CBF
anionic	CBF	CBF	CBF

Table 2.1: *Type of final foam film, observed for specific polyelectrolyte/surfactant combinations; adapted from Ref.[15]*

tant are carefully chosen to get homogeneous and continuously thinning films. The surfactant concentrations should be below the critical micelle concentration (cmc) and the polyelectrolyte concentration below the overlap concentration (c*), to avoid the occurrence of structural forces. The cmc is the surfactant concentration, where micelles are formed in the bulk solution, while c* corresponds to the polyelectrolyte concentration where polyelectrolytes start to overlap and to form a network. In addition, it is important to choose the composition of both components such that the critical aggregation concentration (cac) is not exceeded, since otherwise, aggregates are formed in the solution that prevent the formation of homogeneous films.

2.2.1 Pure surfactant films

Since symmetric films from pure water are not stable, surfactant is added to the system to stabilise these films. The molecules are adsorbed at the film interfaces, where the hydrophilic headgroup is situated in the water phase, while the hydrophobic alkyl chain is exposed to the gas phase. Depending on the type of the surfactant that is used, the properties of the respective foam films differ. In case of ionic surfactants, the charge of the film surface is determined by the surface coverage and the dissociation degree of the surfactant. With increasing adsorbed amount, the charge at the surface increases, leading to a higher repulsion between the two interfaces. In contrast to that, the dissociation of the surfactant molecules increases the ionic strength and therefore the electrostatic screening, counteracting the above described effect. Usually, the increase of the surfactant concentration leads to thinner and more stable films,[16] indicating that the electrostatic screening dominates the interactions in the film.

In systems with sufficiently high surface charges, CBF are formed since the repulsion between the interfaces prevents the transition to a NBF. Therefore, ionic surfactants form CBFs,[2,16,17] unless the amount of salt that is added to the solution exceeds the surfactant concentration by far. In this case, the screening of the surface charges gives rise to a CBF-NBF transition.

In case of nonionic surfactant, CBFs can be formed as well, only above a certain surfactant concentration, a transition to a NBF can be observed. Since a CBF requires electrostatic repulsion of the two interfaces, which can not be originated from the nonionic surfactant, this indicates charges at the neat air/water interface. These charges are continuously replaced by the surfactant molecules which leads to a NBF at higher concentrations. The origin and the sign of the charged surface have been the subject of controversial debates in literature. Experiments, like droplets or bubbles in an electric field[18] or wetting film studies[8,9] indicate, that the water surface is negatively charged. On the other hand, theoretical calculations predict a positively charged surface,[19] but so far there is no experimental proof for that.

In a study by Radke et al.[20] the problem of origin the surface charge has been investigated. They claim that charge of the air/water interface is negative and that is due to the adsorption of OH$^-$ at the surface.

2.2.2 Likely charged polyelectrolyte/surfactant systems

When likely charged polyelectrolytes and surfactants are mixed, no significant influence on the surface tension can be detected. It is assumed that the adsorbed surfactant molecules at the interface repel the likely charged polyelectrolyte, so that no surface-active complexes are formed. Furthermore, hydrophobic interactions between the hydrophobic backbone of the polyelectrolyte and the alkyl chain of the surfactant can occur, resulting in a depletion of the surfactant from the surface. This is confirmed by a slightly higher surface tension for C_{16}TAB/PDADMAC system compared to the pure surfactant.[15] The disjoining pressure isotherms of the same system (with and without polyelectrolytes) show CBFs. They are very stable and show no transition to a NBF in the investigated pressure range. The film of the mixed system is less stable but has the same film thickness. This indicates that the polyelectrolyte does not act like a simple salt, which would lead to electrostatic screening, but does not increase the osmotic pressure either. However, the addition of polyelectrolytes leads to a reduced stability of the foam films, which indicates that the mobility of the polyelectrolyte chains in the film bulk has a significant influence on the film stability.

2.2.3 Oppositely charged polyelectrolyte/surfactant systems

Fig. 2.4 shows a characteristic surface tension curve for oppositely charged polyelectrolyte/surfactant systems. For these systems, surface tension is usually depicted with fixed polyelectrolyte concentration and varied surfactant concentration. Typical concentration ranges are 10^{-4} to 10^{-2} monoM for the polyelectrolyte and 10^{-6} to 10^{-2} M for the surfactant. At low surfactant concentrations (below 10^{-4} M), the addition of polyelectrolytes leads to the formation of surface complexes that lower the surface tension compared to the pure surfactant. This happens at the csac (critical surface aggregation concentration).[21,22] The driving force for the complexation between oppositely charged polyelectrolytes and surfactants are electrostatic and hydrophobic forces. The head groups of the surfactant molecules are attracted by the charged polymer segments. Additionally, the hydrophobic surfactant tails can interact with the hydrophobic backbone of the polyelectrolyte[23] (cf. Fig. 2.5).

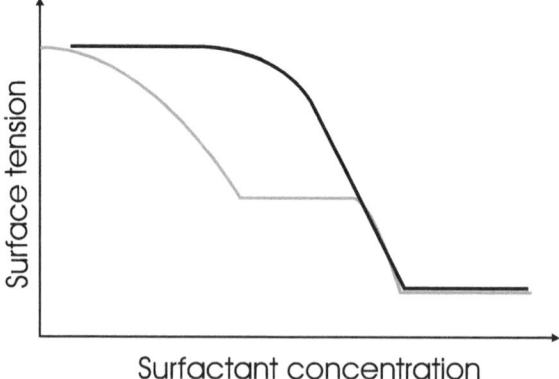

Figure 2.4: *Schematic drawing of the surface tension of pure surfactant (black) and of oppositely charged polyelectrolyte/surfactant mixtures (gray).*

In this concentration regime, only loosely packed monolayers are formed at the interface.[4,24,25]

2.2 Polyelectrolyte/surfactant mixtures in foam films

The distance between the surfactant molecules depends on the degree of charge of the polymer and can be calculated from the Gibbs equation that is applied to the surface tension isotherms. The polymer is linked to the surfactant monolayer with the charged monomer units while the uncharged parts dangle into the bulk solution. The lower the degree of charge, the larger is the distance between the surfactant molecules due to the longer distance between the charged monomer units.[25] In case of $C_{12}TAB$/PAMPS the area per $C_{12}TAB$ molecule is 78 $Å^2$ for a charge fraction of 25% and 100 $Å^2$ for 10% charged monomer units which is much larger compared to 48 $Å^2$ of a surface densely covered with $C_{12}TAB$ molecules.[26,27] The degree of charge has an influence on the thickness of the interfacial layer as well. Highly charged polyelectrolytes such as PSS adsorb flatly at the interface, polyelectrolytes with a lower degree of charge form thicker layers due to the loops that are extended into the solution.[28,29] In a Langmuir-Blodgett film study by Lee et al.[30] it was shown that PSS/$C_{14}TAB$ forms very homogeneous layers at the interface at a surfactant/polyelectrolyte segment ratio of 1. This behaviour depends on the surfactant chain length: shorter chain lengths form layers with holes.

The rigidity of the polyelectrolyte is another important parameter concerning the layer thickness. PSS and PAMPS are rather flexible polyelectrolytes, whereas Xanthan and DNA are more rigid due to their ability to form double helices. Ellipsometry measurements show that they form thicker and denser layers[28] which can be interpreted as the helices adsorbing flat at the surfactant monolayer. In general, the adsorption process is very slow, especially in the low concentration regime[21,31] so that it is important to give enough time to the system for equilibrium.

When the surfactant concentration is further increased, the surface tension reaches a plateau. This plateau is characterised by the cac at which aggregates are formed in the bulk. These aggregates turn the solution turbid when they reach a certain dimension.[32] In this concentration regime, the polyelectrolytes are hydrophobised by the surfactant and the hydrophobic domains aggregate, so the the polyelectrolyte/surfactant complexes precipitate out of the solution. At surfactant concentration above the cmc, the aggregates are described as polyelectrolyte chains that are decorated with surfactant micelles.

The second break point in the surface tension graph in Fig. 2.4 is the cmc. Above this point the surface is densely covered and the surface tension does not change anymore. It is assumed that most of the polymer is redissolved within the bulk in this regime of the surfactant concentration.

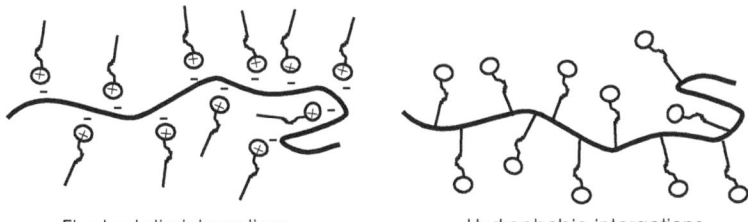

Electrostatic interactions   Hydrophobic interactions

Figure 2.5: *Schematic drawing of the possible interactions between polyelectrolytes and surfactant molecules.*

The cac depends on many parameters like surfactant chain length, polyelectrolyte concentration, or degree of charge. It decreases with increasing surfactant chain length due to the stronger hydrophobic interactions between the two compounds.[33,34] Polyelectrolytes with a high degree of charge interact stronger with the surfactant than those with only a few charged monomer units. Therefore the cac decreases with increasing polymer charge density.[35] The chain length of the polymer has no influence on the cac but in case of PSS a length of at least 20 monomer units is needed to show polymer behaviour.[36]

C_{12}TAB/PAMPS(25%) is an example for a rather hydrophilic system due to the short aliphatic chain of the surfactant and the quite hydrophilic polyelectrolyte. The interaction between both compounds is rather weak due to the lack of hydrophobic interactions and weak electrostatic interactions due to the low degree of polymer charge.[27]

In case of C_{16}TAB/PSS the hydrophobic interaction is so strong due to the long aliphatic chain and the hydrophobic polymer backbone that the charges of both compounds are directed towards the solution and the complexes become hydrophilic and water soluble. That leads to a shift in surface tension to higher values compared to those of the pure surfactant. For combinations of PSS with either C_{12}TAB or C_{14}TAB the surface tension is always lower than for the respective pure surfactant system.[36]

In contrast to PSS, mixtures of C_{16}TAB with PAMPS(25%) and C_{12}TAB/PAMPS(25%) reduce the surface tension with respect to the pure surfactant system.[25] PAMPS shows much weaker hydrophobic interactions with C_{16}TAB than PSS, which makes the complexes much less hydrophilic.

The concentration of the polymer has a more complex impact on the point of aggregation: in general the surface tension at a fixed polyelectrolyte concentration decreases monotonously or remains constant with increasing surfactant concentration. However, in the dilute concentration regime of the polyelectrolyte (far below 10^{-3} M) another phenomenon occurs: when the surfactant concentration has reached the described plateau (close to cac) the surface tension starts to increase again and finally collapses on the isotherm of the pure surfactant. This is more pronounced for surfactants with a long hydrophobic tail. For instance, fully charged PAMPS shows this non–monotonous behaviour in the low concentration regime in combination with C_{16}TAB, whereas the maximum becomes smaller with C_{14}TAB and almost vanishes with C_{12}TAB. So far, this has only been observed for highly charged polyelectrolytes, while 25 % charged PAMPS in combination with C_{16}TAB does not show this non–monotonous behaviour. The starting point of this rise in surface tension is shifted to lower surfactant concentrations when the amount of polymer is reduced. Taylor *et al.*[37] propose a complex ordering of the surface complexes including the formation of multilayers in this concentration regime. These structures are described as several layers with different polyelectrolyte/surfactant compositions, usually a surfactant monolayer with up to 8 mixed layers underneath.[24,37,38] These results are supported by neutron reflectivity experiments. Surface tension measurements indicate a long equilibration time (more than one hour) of the surface in this concentration regime, which might be a hint for the formation of a multilayer at the surface.

Above the cac it is not possible to form homogeneous foam films due to the aggregates that are trapped in the film.

The surfactant/polymer ratio has an important influence on the surface tension as well. At ratios close to 1, Monteux *et al.*[39] observed very hydrophobic complexes in their surface tension measurements for C_{12}TAB/PSS system. These findings are supported by surface rheology studies[40–43] where the surface shows high elasticity, which indicates a large amount of material at the interface. It was even proposed that surface tension is just a question of polymer/surfactant ratio[36] due to the fact that the surface tension was constant when plotted versus the ratio.

The surface tension of C_{12}TAB/PSS mixtures depends on PSS concentration. With decreasing polymer concentration the cac decreases. These findings are in contrast to the results for C_{12}TAB/PAMPS(25%), where all surface tension measurements collapse on one curve in a polyelectrolyte concentration regime of 0.3 to 3 mM.[27] This means that the surface tension is not dependent on the polyelectrolyte concentration in that case. It is not fully clarified, if the difference in charge density or the hydrophobicity of the polymer backbone is responsible for that difference.

The characterisation of the surface complexes has been the subject of many studies in the past years. Concerning foaming and foam film stability, it is hard or even impossible to make predictions only from the surface coverage, *i.e.* surface tension.[44]

2.2.4 Mixtures of nonionic surfactant and charged polyelectrolytes

The addition of positively charged polyelectrolyte like PDADMAC to an aqueous solution of nonionic surfactants like $C_{12}G_2$ has only a minor effect on the surface tension. This indicates that no surface-active complexes are formed in the mixture. A slight decrease in surface tension can be explained by the adsorption of the hydrophobic polyelectrolyte backbone to the surface. As described above, the water/air interface is assumed to be negatively charged due to OH^- adsorption, which could enhance the adsorption of the polyelectrolyte.[15]

The corresponding foam films of these mixtures show a NBF transition at rather low disjoining pressures ($\approx$ 800 Pa). The formation of a CBF can be explained by the electrostatic repulsion of the precharged air/water interface. Above a certain pressure, the cationic polyelectrolyte screens the negative charges at the surface and leads to the formation of a NBF. After the transition, the newly formed NBF breaks after a short period of time. It is very unstable compared to the NBFs induced by the addition of salt or by the increase of surfactant concentration in the case of nonionic surfactant. This could be due to the fact that the surface coverage is less dense or that the fluctuating polymer chains disrupt the ordering in the film, which leads to the film rupture.

When the positively charged polyelectrolyte is exchanged by a anionic polymer like PSS, no NBF transition can be observed before the film rupture at $\approx$ 6000 Pa. This is a hint for a stronger electrostatic repulsion in case of $C_{12}G_2$/PSS compared to $C_{12}G_2$/PDADMAC due to the addition of negative charges to the system.

2.2.5 Stratification phenomena

One of the main topics in the investigation of foam films containing polyelectrolytes is the discontinuous thinning of a film, the stratification process. This stepwise thinning of the foam film is observed below the cac in the semidilute concentration regime of the polyelectrolyte.[45] Stratification occurs due to an oscillation of the disjoining pressure in the film and is assumed to be originated in a transient polyelectrolyte network that is formed in the film core above c^*.[46–49] It is assumed that layers of the network are pressed out of the film when the applied pressure is increased. This is an irreversible process, since it is not possible to go back to another branch of the oscillation when the pressure is reduced. The steps in thickness follow a power law that scales with $\Delta h \propto c^{-1/2}$ for linear polyelectrolytes and $\Delta h \propto c^{-1/3}$ for branched polymers like PEI.[50] In case of a linear polymer, this corresponds to ξ the mesh size of the polymer network that is formed in the bulk.[22,51,52]

The stratifications are affected by the salt content of the solution and the degree of charge of the polyelectrolyte. The addition of salt reduces the disjoining pressure at which the stratification is induced and, above a threshold concentration, suppresses the occurrence of the steps in the film. In the case of charged polyelectrolytes, the step size remains constant down to the Manning threshold. However, for polyelectrolyte with a lower degree of charge, the step size increases and all steps take place simultaneously at a very low disjoining pressure.[46,53,54] Nonionic polymers induce no stratification at all.

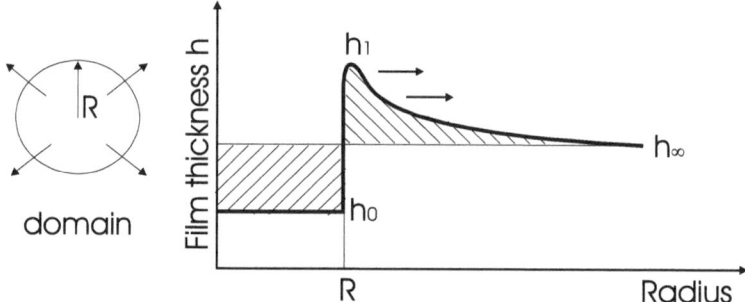

Figure 2.6: *Sketch of a growing domain of the radius R; the film thickness equals h_0 inside the domain and h_∞ at infinity; the film tension difference between the inside and the outside results in a rim with height h_1; material transport is marked by black arrows.*

The formation of the network of polymer chains has been widely discussed in literature. The question has been arisen if it is really a network that is formed in confinement or if there rather is a layering of the polymers with a pronounced alignment parallel to the surface.[51,55,56] Fluorescence measurements[57] of pyrene-labeled PAA[58] have shown that there indeed is a layer wise arrangement but with a distance between the layers that corresponds to a random polymer network in the respective bulk phase.

Backbone rigidity also plays an important role in the stratification process. For flexible polyelectrolytes the force oscillation can be observed as long as the velocity of the two approaching surfaces is not too fast.[56] More rigid polymers show stratification only when the viscosity of the solution is large enough so that the network has time to adjust, otherwise no stepwise thinning can be observed.

The choice of the surfactant has no detectable influence on the structuring of the polyelectrolyte chains within the film core, which leads to a fixed period of force oscillation at a certain polyelectrolyte concentration independent on the surfactant type.[15,59] This means that the interaction between the polyelectrolytes and surfactant molecules can be neglected with respect to their effect on structural forces. For example, the disjoining pressure isotherms of free-standing aqueous films containing the polycation PDADMAC in combination with either nonionic $C_{12}G_2$ or positively charged $C_{16}TAB$ both show stratification.[15] The results confirm that the surfactant has no influence on the step size, and therefore, on the structuring of the polyelectrolytes within the film. Beside the charge also the elasticity of the interfaces has no effect on the structural forces. The force oscillations in films of polyelectrolyte solutions were not only measured in foam films, but also in wetting films[60] and between two solid interfaces in an AFM.[55,61] The period of the pressure oscillation remains constant in all cases.

In contrast to that, the interactions between polyelectrolyte and surfactant do affect the velocity of the stratification. The stratification process starts with small discontinuities in the film thickness, visible as small dark dots spreading over the hole film. The velocity of the growth of these domains depends on the boundary conditions of the surface. When the polyelectrolyte is linked to the surface, the domain growth is much slower than in the case of a depleted interface.[62,63] The reasons for this could be that the polyelectrolytes chains dangling from the surface slow down the domain growth process. The driving force of this domain growth is the difference in film tension $\Delta\sigma$ between the two film parts. The film tension in the inner part of the domain is smaller than that of the thicker film,[64] and as the system favours the lower energy state of the inner part the domains are enlarged. $\Delta\sigma$ is assumed to result in an increase in the film thickness of the rim surrounding the domain so that h_1 is a material constant (*cf.* Fig. 2.6).

3 Experimental section

3.1 Materials

3.1.1 Surfactants

The cationic surfactants tetradecyl trimethyl ammonium bromide (C_{14}TAB) and dodecyl trimetyl ammonium bromide (C_{12}TAB) were both purchased from Sigma-Aldrich (Steinheim, Germany) and recrystallised at least 3 times from acetone with traces of ethanol. All solutions were prepared from Milli-Q water (resistivity 18.2 M Ω cm^{-1}) with a surfactant concentration of 10^{-4} M unless it is stated otherwise. This concentration is well below the cmc of both surfactants (3.5×10^{-3} M for C_{14}TAB and 1.5×10^{-2} M for C_{12}TAB).

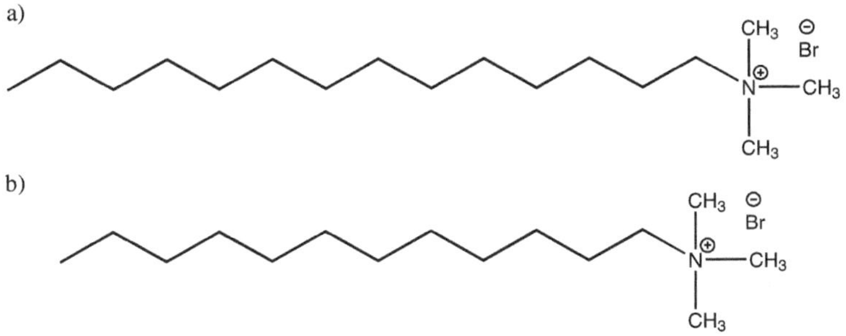

Figure 3.1: *Chemical structure of a) C_{14}TAB and b) C_{12}TAB.*

The anionic surfactant sodium decyl sulfonate ($C_{10}SO_3$) was purchased from Sigma-Aldrich and recrystallised 3 times from ethanol with water. It was used in a concentration regime of 10^{-6} to 10^{-1} M and has a cmc of 4×10^{-2} M.

3.1.2 Polyelectrolytes

The anionic poylelectrolytes poly(acrylamido methyl propanesulfonate) sodium salt (PAMPS) and poly([tris (hydroxymethyl)methyl]acrylamide *co* acrylamido methyl pro-panesulfonate)

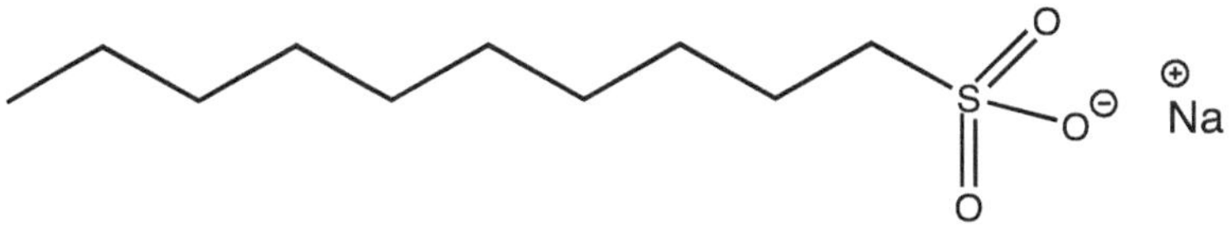

Figure 3.2: *Chemical structure of $C_{10}SO_3$.*

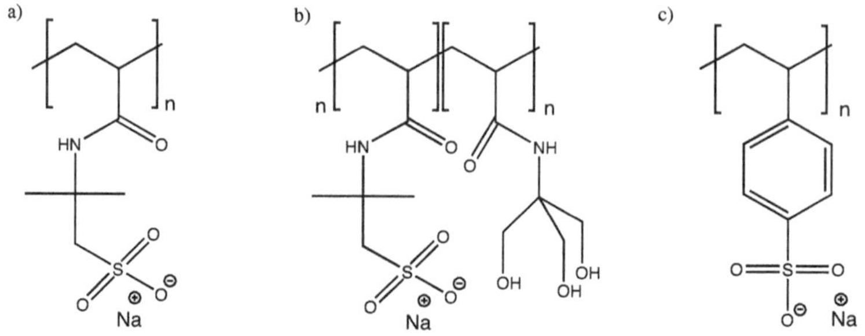

Figure 3.3: *Chemical structure of a) PAMPS, b) (PTRIS-co-AMPS) and c) PSS.*

sodium salt (PTRIS-*co*-AMPS) were prepared by an aqueous free-radical polymerisation.[65] The first was synthesised with a nominal degree of charge of 100 % while the latter has a nominal charge fraction of 60 %. In this thesis it is referred to as PAMPS 60%. Both polyelectrolytes were received from Sebastien Garnier and André Laschewsky (IAP, Universität Potsdam) and were used without further purification.

Poly(styrene sulfonate) sodium salt (PSS) was purchased from Aldrich (Steinheim, Germany). In this work, PSS was used with different molecular weights, namely with MW = 70000, 13000, and 4300 g/mol. This corresponds to a polyelectrolyte chain length of 340, 60, and 20 monomer units, respectively, so that they are referred to as PSS, PSS60 and PSS20. Prior to use, the polyelectrolytes were purified with an ultra filtration cell with a cut-off membrane of 20000 Da for PSS and a cut-off membrane of 1000 Da for PSS60 and PSS20, respectively, to remove all low weight impurities.

The polyelectrolyte concentration was varied between 10^{-5} to 10^{-3} monoM. All given concentrations are related to the respective monomer concentrations.

3.1.3 Salts and Monomers

The monomers sodium styrene sulfonate (NaSS) and acrylamido methyl propanesulfonic acid (AMPS) were purchased from Sigma-Aldrich (Steinheim, Germany) and used as received. To get the sodium salt of AMPS, the acidic stock solution was neutralised with an appropriate amount of 0.1 M NaOH. Sodium chloride (NaCl) from Merck (Darmstadt, Germany) was roasted at 500 °C to remove all organic impurities.

3.2 Methods

3.2.1 Thin Film Pressure Balance (TFPB)

A common method to measure disjoining pressure isotherms is the Thin Film Pressure Balance (TFPB). It was first developed by Mysels and Jones,[66] later improved by Exerowa[67,68] and allows the formation of horizontal, free-standing foam films.

In the original model of a TFPB, the film is formed in a glass ring where the liquid is sucked out through a small hole (the so-called Scheludko cell). This method has strong limitations regarding the applicable disjoining pressure range due to the low capillary entry pressure of the

3.2 Methods

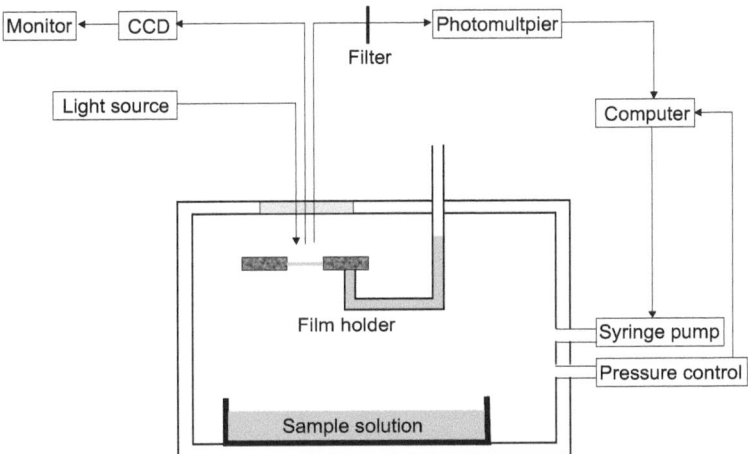

Figure 3.4: *Sketch of a Thin Film Pressure Balance.*

hole. Furthermore, the hole connecting the film to the bulk liquid in the capillary prevents an equal drainage of the film. These limitations are overcome by the development of the porous plate technique, proposed by the groups of Mysels and Exerowa.[69,70] In this technique, the film is formed in a hole with a diameter of ≈ 1 mm, that is drilled into a porous glass disc with a pore size of 10-15 μm (porosity P4), which allow capillary entry pressures up to 10000 Pa.

Fig 3.4 shows the setup of the home-built TFPB used in this thesis. It consists of a stainless steal measuring cell that is equipped with a quartz glass window to monitor the film and temperature control. The film holder is placed in the sealed cell and connected to the ambient reference pressure. A reservoir of the sample solution is located in the cell to provide a saturated vapor atmosphere during the measurement. The measuring chamber is placed on a vibration isolated table. The applied pressure is adjusted via a motor-driven syringe pump and controlled by a computer. To minimize the error during the measurement, two differential pressure transducers (DPT)(MKS Instruments, München, Germany) are used, one for low pressures up to 1000 Pa and the second for measurements between 1000 and 10000 Pa. Since the DPTs have an error of 0.3 % of their full range, this setup allows a more accurate measurement in the low disjoining pressure regime. The increase of the gas pressure in the chamber leads to a flow of the excess liquid from the droplet through the pores of the glass disc, and the symmetric foam film is formed. At equilibrium, the difference between the inner and the outer pressure is balanced by disjoining pressure Π. The film is observed with an optical microscope (Nikon, Düsseldorf, Germany) and illuminated by cold filtered white light to avoid additional energy input in the system. The reflected light beam is split into two parts: one part is used to monitor the film with a CCD video camera (Pulnix Deutschland GmbH, Alzenau, Germany), while the other part is sent through a narrow band-pass filter ($\lambda = 632$ nm) and amplified by a photomultiplier. From the measured intensity of the reflected light, the film thickness is calculated.

Disjoining pressure

At equilibrium, the disjoining pressure Π is balanced by the capillary pressure and is directly determined from the difference between the pressure in the film P_{film} and the pressure of the liquid reservoir P_l:

$$\Pi = P_c = P_{film} - P_l \qquad (3.1)$$

Since the film has planar surfaces (infinite radius of curvature), the film pressure P_{film} equals the gas pressure P_g in the cell:

$$\Pi = P_g - P_l \qquad (3.2)$$

P_l, the pressure of the bulk liquid is determined by the external reference pressure P_r, the capillary pressure and the hydrostatic pressure of the liquid in the attached glass capillary:

$$\Pi = P_g - P_r + \frac{2\gamma}{r} - \Delta\rho g h_c = \Delta P + \frac{2\gamma}{r} - \Delta\rho g h_c \qquad (3.3)$$

ΔP, the difference between the pressure in the chamber and the reference pressure and is directly measured by the differential pressure transducer. The capillary pressure is determined by γ, the surface tension of the sample solution and the inner radius of the capillary tube r. The hydrostatic pressure is calculated from the height of the liquid column h_c above the level of the film. $\Delta\rho$ is the density difference between the solution and the gas phase and g the gravitational acceleration.

Determination of the film thickness

To measure the thickness of a foam film, the film is illuminated with white light and the intensity of the reflected light is recorded. The incoming light is reflected at the upper and the lower film interface and the two waves interfere with a phase difference that is correlated to the film thickness. The interferometric method developed by Scheludko[71,72] scales the intensity of the interfered light with the interference minimum I_{min} and maximum I_{max}. In case of a symmetric film, the film thickness is calculated according to the following equation:

$$h_{eq} = \frac{\lambda}{2\pi n_s} \arcsin \sqrt{\frac{\Delta}{1 + (4R(1-\Delta)/(1-R)^2)}} \qquad (3.4)$$

with

$$\Delta = \frac{I - I_{min}}{I_{max} - I_{min}} \qquad (3.5)$$

and

$$R = \frac{(n_s - 1)^2}{(n_s + 1)^2} \qquad (3.6)$$

where λ is the wave length of the used interference filter ($\lambda = 632$ nm), n_s the refractive index of the sample solution ($n_s = 1.33$) and I the instantaneous light intensity. I_{max} and I_{min} are determined during the formation of the film and after film rupture, respectively. At the maximum, the foam film has a thickness of approximately 118 nm, depending on the λ of the interference filter. With decreasing film thickness, the film gets darker due to the increasing destructive interference between the two reflected light waves.

The equivalent thickness in eq. 3.4 is slightly thicker than the true film thickness, h, because

3.2 Methods

the adsorption layers at the film interfaces have a higher refractive index than the aqueous core.[3] It is possible to correct h_{eq} by using a three-layer model, taking the different refracting indices into account. Since the difference in h and h_{eq} is relatively small for CBFs (below 5 %), this correction is only important for very thin films.

Measuring procedure

The used film holders are home-made from porous glass discs (Robu, Hattert, Germany, diameter = 30 mm, porosity 4) with a hole of 1 mm diameter and a glass capillary (inner diameter = 3 mm) that was attached to it to connect the system to the outer reference pressure. Prior to the measurements, the film holder was cleaned 10 times with ethanol and Milli-Q water and boiled in hot water for several hours. Each film holder could only be used for one sample system due to adsorption of the charged molecules to the glass surface. The polyelectrolyte/surfactant and electrolyte/surfactant solutions were freshly prepared from a stock solution before each measurement. All glassware, except film holders, was cleaned in a 5 % Q9 solution over night and thoroughly rinsed with Milli-Q water.

Directly before the measurement, the film holder was boiled in hot water for 30 min, dried in a nitrogen stream, rinsed 5 times with the respective sample solution, and immersed into the solution for 2 h to saturate the pores. The immersed film holder was placed in the sealed measuring cell to get a saturated atmosphere. 30 min before starting the measurement, the holder was pulled out of the solution to give the surfaces of the droplet time to equilibrate. In this work, equilibrium measurements have been carried out. For each data point, the intensity of the reflected light was recorded until it was constant for 20 min. Each experiment was reproduced at least 3 times at a temperature of 23 °C.

3.2.2 Surface Characterisation

Surface Tension Measurements

Surface tension measurements have been performed to get more information about the surface properties of the film. Since it is not possible to measure the surface tension directly at the film interface, a single air/water interface has been investigated. It has been assumed that the changes of the surface composition[73] that occur during the approach of the two air/water interfaces upon film formation are only minor, so that the results of the surface tension measurements can be transferred to the film surfaces.

The surface tension of liquids originates from a difference in the energy state between the surface molecules and those in the bulk. Molecules that are located at the surface are only partly surrounded by other molecules, so that they are exposed to a higher energy compared to the bulk molecules. Since it is more favourable for the system to be in a lower energy state, the liquid tends to minimize the surface area, so that more molecules are situated in the bulk solution and less at the surface.[74] Hence, work is required when a new surface with the area A is created:

$$w = \int_0^A \gamma dA = \gamma A \tag{3.7}$$

with the surface tension γ being the constant of proportionality.[75]

The addition of even small amounts of surface-active molecules reduces the surface tension significantly. To get more information about the adsorption of surface-active components, the surface concentration Γ can be calculated from the surface tension isotherm by using the Gibbs equation:

$$\Gamma = -\frac{1}{RT}\frac{d\gamma}{d\ln c} \qquad (3.8)$$

It works reasonably well for surfactant molecules, for which adsorption is reversible.[76]

In this work, the surface tension was measured with a K11 Tensiometer from Krüss (Hamburg, Germany) using the du Noüy ring. This method is based on force measurements of a metal ring made of a thin Pt/Ir wire (with radius R), that is immersed into the solution and slowly pulled out. The surface tension is obtained from the force that is necessary to detach the ring from the surface of the liquid.[13]

$$\gamma = \frac{F}{4\pi R} \qquad (3.9)$$

The force F is corrected by the weight of the lamella that is attached to the ring. Furthermore, it is important to consider twice the length of the wire, since the liquid touches the inner and the outer part of the ring. This is reflected in the denominator in eq. 3.9.

All experiments were carried out in a Teflon trough with a diameter of 5 cm at 23 °C. The trough with the respective sample solution was tempered in the tensiometer for 15 min, and a lamella was pulled out of the solution for 20 min to let the polyelectrolyte/surfactant complexes adsorb at the surface. The surface tension values given in this study are the average of at least 3 measurements.

Dynamic Surface Elasticity Measurements

To get a deeper insight into the rheological properties of the surface, surface elasticity measurements were performed, using the oscillating drop method. This surface rheological approach considers the response of the interfacial tension to the dilational deformation of the adsorbed layer (expansion and contraction of the surface). Therefore, the surface area of a drop is changed by generating harmonic oscillations while simultaneously the surface tension is recorded.

The surface tension of the droplet can be calculated by the analysis of the shape of the pendant drop. The shape is determined by two competing forces: The gravity that elongates the droplet and the surface tension that counteracts the gravity to minimize the surface of the drop.

The oscillating drop method became accessible when new electronic video cameras were developed to get images of the pendant drop from which the drop profile coordinates could be extracted and then fitted by the Gauss-Laplace equation:

$$\Delta P = \gamma(\frac{1}{R_1} + \frac{1}{R_2}) \qquad (3.10)$$

The curvature is adjusted such that the difference in the pressure between the two phases is balanced by the capillary pressure. In the absence of any external forces other than gravity, this may be expressed by the following expression:

$$P^0 + \Delta\rho g z = \gamma(\frac{1}{R_1} + \frac{1}{R_2}) \qquad (3.11)$$

where P^0 is the capillary pressure at the drop apex, $\Delta\rho$ is the density difference between the internal and the external phases, g is the gravitational acceleration, and z is the height of the

3.2 Methods

droplet measured from the reference plane. R_1 and R_2 are the principal radii of curvature.[77]

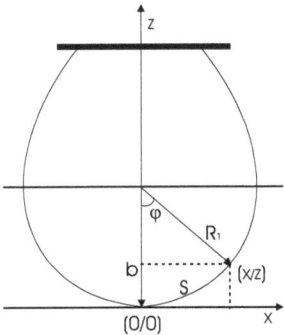

Figure 3.5: *Sketch of a pendant drop.*

In the special case of axis-symmetric droplet, the model shape is calculated from the Gauss-Laplace equation by using a fourth-order Runge-Kutta integration algorithm. This is represented in a set of three first-order differential equations:

$$\frac{dx}{dS} = \cos(\varphi) \quad (3.12)$$

$$\frac{dz}{dS} = \sin(\varphi) \quad (3.13)$$

$$\frac{d\varphi}{dS} = \frac{1}{R_1} = \pm\frac{\Delta\rho g z}{\gamma} + \frac{2}{b} - \frac{\sin(\varphi)}{x} \quad (3.14)$$

where S is the arc length to an arbitrary point of the profile, b the radius of curvature at the apex and φ is the normal angle[78,79] (cf. Fig. 3.5).

The dilational deformation of the droplet causes a change of the surface concentration: Expansion leads to dilution of the surface layer and hence, to an increase of γ, while the contraction of the interface results in a decrease of the surface tension. Additionally, molecular exchange with the adjacent bulk phase can take place.

E is defined by the following equation:[80]

$$E = \frac{d\gamma}{d\ln A} \quad (3.15)$$

If a relaxation process occurs during the compression, a compression viscosity can be introduced additionally to the elasticity. In the case of sinusoidal deformation and small amplitudes, it is related to the imaginary part of the complex modulus.[1]

$$E(\omega) = E_r + iE_i = E_r + i2\pi\nu\eta \quad (3.16)$$

The surface elasticity modulus is calculated from the amplitude ratio of the oscillating surface tension and surface area, whereas the phase shift between the two determines the dilational viscosity.[77]

The surface dilational modulus E is obtained as a function of the frequency ν and the sinusoidal oscillations of the surface A. At a given frequency, A is described by the following equation:

$$A = A^0 + \tilde{A}\sin 2\pi\nu t \qquad (3.17)$$

where A^0 is the original undisturbed surface area and $\tilde{A}$ the amplitude of the area oscillation. The perturbation produces a harmonic response of the surface tension γ with the same frequency:

$$\gamma = \gamma^0 + \tilde{\gamma}\sin(2\pi\nu t + \phi) \qquad (3.18)$$

with γ^0 being the surface tension of the undisturbed surface, $\tilde{\gamma}$ the amplitude of the surface tension oscillations, and ϕ the phase shift between sinusoidal disturbance and response. This leads to a complex modulus that can be expressed in terms of measurable quantities[81]

$$E = A^0 \frac{\tilde{\gamma}}{\tilde{A}}\cos\phi + iA^0\frac{\tilde{\gamma}}{\tilde{A}}\sin\phi \qquad (3.19)$$

For this technique, the generated harmonic oscillations should not exceed 1 Hz. These low frequencies guarantee slow perturbations, so that the drop keeps its Laplacian shape.[77,82]

Measuring procedure

The surface elasticity measurements were performed at a PAT1 (Sinterface Technologies, Berlin, Germany) in cooperation with Reinhard Miller from the Max Planck Institute for Colloids and Interfaces. Before each measurement, the apparatus was thoroughly rinsed with ethanol and Milli-Q water, and the surface tension of water was checked to exclude the presence of impurities. The pendant drop was created at the tip of a capillary by a computer-driven dosing system. The drop was placed in a closed cuvette with a small reservoir of the sample solution at the bottom to prevent evaporation. Drop images were acquired with a CCD video camera and from these images, the surface tension was calculated by the drop shape analysis. After droplet formation, the surface was left to equilibrate for at least 2 h. The harmonic oscillations of the drop surface area were created with the computer-controlled dosing system and both, surface area A and surface tension γ were monitored as a function of time. The frequencies of the oscillation were varied between 0.005 and 0.1 Hz with at least 6 oscillations per frequency. All measurements were carried out at room temperature. After data acquisition, a Fourier transformation was performed and the surface elasticity was derived by the described method.

4 Effect of surface charge on foam film stability*

Abstract

The present work deals with the control of the stability of ionic surfactant (C_{14}TAB) foam films by the addition of oppositely charged polyelectrolytes (PAMPS). In the two cases at low and high polyelectrolyte concentrations, a common black film (CBF) is formed due to an electrostatic repulsion. At high polyelectrolyte concentrations, it was assumed that a charge reversal takes place at the film surfaces. But what happens around the nominal isoelectric point (IEP), where the net charge of the polyelectrolyte/surfactant complexes is close to zero? Is a Newton black film (NBF) formed or does the film break? Disjoining pressure isotherms show a strong reduction in stability close to the IEP. The comparison between surface tension and elasticity measurements and disjoining pressure isotherms leads to a surprising conclusion: The stability of foam films seems not to be dominated by the net charge of the polyelectrolyte/surfactant complexes at the surface, which was the former hypothesis, but rather by the overall excess charge of these complexes within the film.

4.1 Introduction

The properties of foams are of interest for many industrial applications like enhanced oil recovery and in personal care products and are therefore the subject of many studies.[17] In some processes, foam is desired and in others it should be avoided, which shows the high impact of the control of foam stability. The addition of polymers plays an important role *e.g.* for surface protection and decalcification processes. In order to control and manipulate the properties of a foam it is essential to understand the behaviour of the single building blocks, the so-called foam films.

One can distinguish between two different types of films: The common black film (CBF) that is mainly stabilised by electrostatic forces and that has a thickness of 10-100 nm and the Newton black film (NBF) which is dominated by steric forces. A NBF consists only of the two layers of absorbed molecules including their hydration shell. Hence, its thickness corresponds to twice the surfactant lengths, *i.e.* about 4 - 5 nm for surfactants of low molecular weight. All other liquid is pressed out of the film.[7]

The present chapter focuses on oppositely charged polyelectrolyte/surfactant mixtures, namely on cationic surfactants and negatively charged polyelectrolytes. When the two compounds are mixed, highly surface-active complexes can occur. They have been investigated by means of surface tension measurements,[25,27] neutron[24,83,84] and X-ray reflectivity,[28,85] ellipsometry[26,39] and surface elasticity.[31,40,86]

Adding polyelectrolytes changes the properties of a foam film significantly. Depending on the

---

* Similar content has been published in : *No Charge Reversal at Foam Film Surfaces after Addition of Oppositely charged Polyelectrolytes?*, N. Kristen, V. Simulescu, A. Vüllings, A. Laschewsky, R. Miller and R. v. Klitzing, *J. Phys Chem. B*, **2009**, *113*, 7986-7990

charge combination of both compounds different types of films are formed. So far, NBFs were only observed for the combination of nonionic surfactant and cationic polyelectrolyte.[15,87] The slightly negatively charged air/water interface[8] attracts the polyelectrolyte and induces the formation of a NBF, but no surface complexes are formed in this case, which was shown by surface tension measurements.

Likely charged polyelectrolyte/surfactant systems form CBFs due to strong repulsion between the molecules.[49,51,88] The same effect is observed for oppositely charged components,[45–47] which is counterintuitive. At both low and high polyelectrolyte concentrations CBFs are formed due to the electrostatic repulsion. So far, it has been assumed that the addition of oppositely charged polyelectrolytes below the cac leads to adsorption of polyelectrolytes and therefore first to a reduction in net charge of the film surface followed by a charge reversal. Then, the surface charge has the same sign of charge as the polyelectrolyte, which leads again to a strong repulsion between polyelectrolyte and the film surfaces and therefore to the formation of a CBF.

The aim of this work is to study the influence of the surface charge on the stability of the foam films. Especially, the stability close to the nominal IEP in a film containing oppositely charged surfactants and polyelectrolytes is addressed. At very low surface net charge, which is assumed to be close to the nominal IEP, two scenarios are possible: The formation of a NBF or the destabilisation of the film. Starting with pure cationic surfactant films, the film surfaces are positively charged.[89] To get stable films in this regime, it is important to choose a surfactant with a sufficiently long hydrophobic chain. For the C_nTAB series this is C_{14}TAB or larger.[17]

To control the surface coverage it is essential to adjust the concentrations such that the polyelectrolyte concentration is below the critical overlap concentration (c*) and the surfactant concentration below the critical micelle concentration (cmc). Otherwise the effects of surface charges would be superimposed by stratification of the film. In the case of a stratification, the film drains stepwise due to an oscillation of the disjoining pressure.[15,45,47,48,51] This oscillation is induced by a transient network of the polyelectrolyte chains in the film core and by the layering of micelles, respectively. This phenomenon corresponds to the findings in polymer bulk solutions[88] and can be generalised for many different types of polyelectrolytes. It does not depend on the choice of the surfactant and can also be found between two solid surfaces.[52]

In addition, above a certain concentration, the so-called critical aggregation concentration (cac), polyelectrolyte and surfactant start to form pronounced bulk aggregates which lead to inhomogeneous films.[22,45] To avoid this, it is crucial that the concentrations of both compounds are adjusted in a way that they are below the cac in the whole concentration regime.

To get a deeper insight into the behaviour of the films and the complex formation in the solution bulk and at the interfaces, disjoining pressure isotherms, surface tension and elasticity were measured.

4.2 Results

We have investigated polyelectrolyte/surfactant solutions of a fixed surfactant concentration and variable polyelectrolyte concentration. The polyelectrolyte concentration was varied between 10^{-5} M and 10^{-3} M whereas the surfactant concentration was fixed at 10^{-4} M. All polyelectrolyte concentrations given in this paper refer to the concentration of monomer units and are below the cmc, the cac and c*.

Fig. 4.1 shows the disjoining pressure isotherms of C_{14}TAB solutions at different PAMPS concentrations. At low disjoining pressure the film thickness is about 100 nm. The stability of the isotherms below the nominal IEP at a polyelectrolyte concentration of 10^{-4} M decreases with

4.2 Results

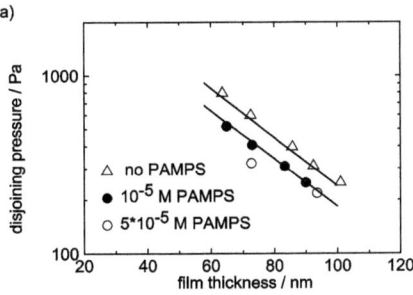
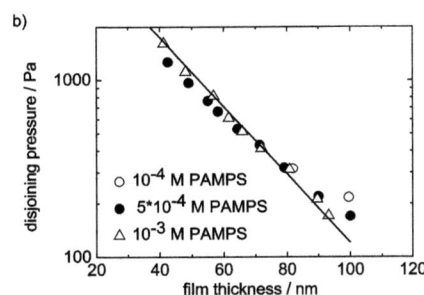

Figure 4.1: *Disjoining pressure isotherms for $C_{14}TAB$ solutions with PAMPS for different polyelectrolyte concentrations: a) below the nominal IEP of 10^{-4} M; b) at and above the nominal IEP; the solid lines correspond to simulations at constant potential.*

PAMPS conc. [monoM]	Ψ_0 [mV]	κ^{-1} [nm]	I [M]
no PAMPS	78	29.3	1.0×10^{-4}
1×10^{-5}	53	29.0	9.0×10^{-5}
5×10^{-4}	75	21.8	1.7×10^{-4}
1×10^{-3}	82	16.0	3.0×10^{-4}

Table 4.1: *Summary of the surface potentials Ψ_0 from the simulation of the disjoining pressure isotherms of PAMPS/$C_{14}TAB$ films; the Debye length κ^{-1} is calculated from a fit of the experimental data with a exponential decay function of first order, the ionic strength I is derived from the simulation of the surface potential.*

increasing PAMPS concentration from a rupture pressure of 900 Pa for the pure surfactant film to 300 Pa at a concentration of 5×10^{-5} M PAMPS. At 7.5×10^{-5} M PAMPS it is not possible to form stable films at all. With further increase of the polyelectrolyte concentration the films get even more stable (up to 1600 Pa at 10^{-3} M PAMPS).

Due to the number of data points, it was only useful to do simulations for the most stable films at PAMPS concentrations of 0, 10^{-5}, 5×10^{-4} and 10^{-3} M. The simulations were done at constant potential with the PB program version 2.2.1,[12] solving the non–linear Poisson–Boltzmann equation. The results are summarised in Table 4.1. The value for the ionic strength of 3.0×10^{-4} M in case of a PAMPS concentration of 10^{-3} M indicates that every third monomer unit of the polyelectrolyte is dissociated which is in good agreement with the Manning concept of counterion condensation. At a PAMPS concentration of 10^{-5} M, the ionic strength might be dominated by the C_{14}TAB concentration of 10^{-4} M, which is reflected in a similar value for the ionic strength as for the pure C_{14}TAB film.

To illustrate the stabilities of the disjoining pressure isotherms, Fig. 4.2 shows the maximum pressure that can be applied before film rupture in dependence of the polyelectrolyte concentration. The point of destabilisation ($\Pi_{max} = 0\ Pa$) is at a PAMPS concentration of 7.5×10^{-5} M, *i.e.* slightly below the nominal IEP and splits the graphic into two parts: below this point the film stability decreases, and above it increases again, even above the stability of the pure C_{14}TAB film.

In Fig. 4.3 the surface tension of the corresponding polyelectrolyte/surfactant solutions is

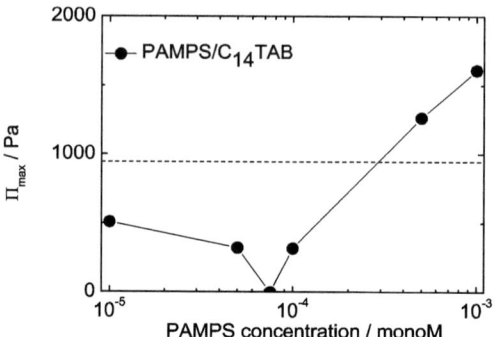

Figure 4.2: *Stability of $C_{14}TAB/PAMPS$ films; maximum disjoining pressure Π_{max} before film rupture versus polyelectrolyte concentration; the dashed line corresponds to the stability of the pure surfactant film at 10^{-4} M.*

shown. Unlike in former publications[28,36,37] where the surface tension is shown in dependence of the surfactant concentration we focus on measurements at a fixed surfactant concentration and varied the amount of polymer. Obviously, the surface tension strongly depends on the PAMPS concentration and shows a non-monotonous behaviour. For the better understanding we divide the curve into three different concentration regimes. In the first regime below the IEP the addition of even very small amounts of PAMPS strongly affects the surface tension. At the lowest measured concentration of 10^{-5} M the surface tension is already lowered to 55 mN/m and reaches its minimum close to 10^{-4} M. In the second regime just above the IEP, the surface tension increases again and reaches a value close to that of the pure surfactant (about 70 mN/m). In the third regime, the surface tension decreases again with increasing polyelectrolyte concentration.

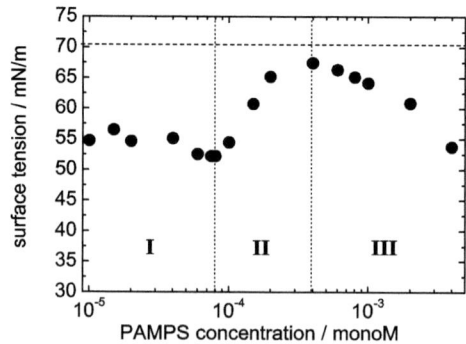

Figure 4.3: *Surface tension of $C_{14}TAB/PAMPS$ solutions with fixed surfactant (10^{-4} M) and variable PAMPS concentration; the dashed line corresponds to the surface tension of the pure surfactant.*

4.3 Discussion

The surface elasticity correlates with the surface tension and again three different regimes can be distinguished (cf. Fig. 4.4). Low surface tension values indicate a high amount of material at the interface which leads to highly elastic surfaces whereas surfaces with less material and hence high surface tension are less elastic. At a PAMPS concentration of 10^{-6} M the elasticity is very low and corresponds to the value of the pure surfactant solution. With increasing polyelectrolyte concentrations the elasticity increases strongly and reaches its maximum of 50 to 60 mN/m close to the IEP. Furthermore, the elasticity range over the frequencies broadens in this concentration regime. Above the IEP, there is a sudden drop to low surface elasticities which is in agreement with the surface tension measurements. After reaching a minimum the elasticity rises again with increasing polyelectrolyte addition.

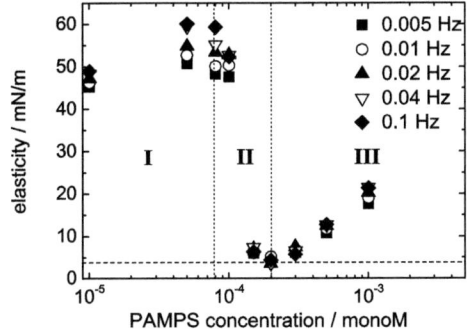

Figure 4.4: *Surface elasticity of $C_{14}TAB$/PAMPS solutions with fixed surfactant (10^{-4} M) and variable PAMPS concentration at different frequencies; the dashed line corresponds to the elasticity of the pure surfactant and the mixture with 10^{-6} M PAMPS.*

4.3 Discussion

The aim of this work was to study the stability of foam films formed from aqueous mixtures of oppositely charged surfactant and polyelectrolytes. The disjoining pressure isotherms of the film, surface tension and elasticity of the respective air/liquid interface show non-monotonous characteristics in dependence of the polyelectrolyte concentration with a reversal point close to the nominal IEP.

4.3.1 Below the nominal isoelectric point.

Below the nominal IEP (10^{-4} M) the surfactant concentration exceeds the concentration of the polyelectrolyte monomer units. The counterions of the polyelectrolyte are exchanged by the surfactant molecules which is energetically favorable for the surfactant due to the resulting decrease of electrostatic repulsion between the surfactant head groups at the surface. The release of the counterions additionally increases their entropy.[27] This exchange leads to polyelectrolyte/surfactant complexes which are hydrophobic and therefore surface-active. This is supported by a decreasing surface tension and an increasing elasticity below the IEP. It is noteworthy that pure surfactant or pure polyelectrolyte solutions within the studied concentration regime show a surface tension similar to pure water. However, below the IEP the kinetics are very slow and hence the equilibration times are very long. This is visible as a frequency dependence of the elasticity. The excess of surfactant suggest that the net charge of the surface is positive. From the isotherms a surface potential of 52 mV was calculated for a PAMPS concentration of 10^{-5} M and the resulting electrostatic repulsion of the interfaces induces the formation of a CBF. The stability of the films decreases towards the IEP.

4.3.2 At the nominal isoelectric point.

At this point the concentration of the polyelectrolyte segments equals the surfactant concentration. As shown in Fig. 4.2, the absence of charge and the resulting electrostatic repulsion of the interfaces leads to the destabilisation of the film and not to the formation of an NBF. The hydrophobic polyelectrolyte/surfactant complexes are located at the surface. Hence, a low surface tension and a high surface elasticity can be observed. However, the solutions do not get turbid and there is no precipitation of aggregates so that we get homogeneous films for all solutions. The results show that the minimum in surface tension, the maximum in surface elasticity and the point of destabilistaion coincide at a polyelectrolyte concentration of 7.5×10^{-5} M. This characteristic point slightly below the nominal IEP, and the shift might indicate a dissociation degree of the surfactant of less than 100%. At a concentration of 10^{-4} M PAMPS stable CBF can be formed, the surface tension increases and the elasticity slightly decreases.

4.3.3 Above the isoelectric point.

In the concentration regime between 10^{-4} M and 10^{-3} M PAMPS there is an excess of polyelectrolyte segments in the solution. Not every charge of the anionic polymer chains can be neutralised by a surfactant molecule so that the net charge of the complex is assumed to be negative. Due to the fixed surfactant concentration less $C_{14}TAB$ molecules decorate one polyelectrolyte chain. Altogether this makes the polyelectrolyte/surfactant complex more hydrophilic and less surface-active as shown by the re–increase in surface tension and the drop of the surface elasticity which both indicate less material at the interface. This phenomenon is also observed by Monteux et al.[33] and Taylor et al.[37] Both quantities go back close to the values of the pure surfactant and the equilibration times are much shorter (no frequency dependence of the elasticity). These findings indicate that most of the polyelectrolyte/surfactant complexes are released from the surface and that a thin surfactant layer and, if at all, a small amount of polyelectrolytes covers the surface.

On the other hand, the foam film stability increases in this concentration regime and gets even higher than for the pure surfactant film. That means that the stability cannot arise only from the described surfactant layer but also from an additional contribution from the negatively charged polyelectrolyte within the film bulk. The calculation of the surface potential from the disjoining pressure simulation gives a value of 83 mV for a PAMPS concentration of 10^{-4} M which is higher than the potential of the pure surfactant film (78 mV). It seems to be an effective surface potential since it does not reflect only the potential at the surface but includes all charges of the system (which are projected onto the surface). In contrast to salts of low molecular weight the polyelectrolyte does not screen the charges but increases the repulsion. Altogether this is an indication that the polyelectrolyte within the film bulk is involved in the stabilisation of the films as well and that the picture of the charge reversal of the interface might be too simple.

As mentioned earlier in the concentration regime above the IEP, the surface tension and the elasticity do not remain constant. So far, the reasons are rather speculative. At PAMPS concentrations higher than 10^{-3} M the critical overlap concentration is reached and a network is formed in the film core[52] which leads to stratification of the film. The stratification is related to an oscillation in polyelectrolyte concentration[57] within the film, which might start at concentrations slightly below the concentration, where stratification can be observed. We suggest that the surfactant is unequally distributed over the network of polymer chains. Those at the borders of the network might carry more surfactant molecules since they are close to the surface. The additional $C_{14}TAB$ molecules hydrophobise the polyelectrolyte/surfactant complexes so that they get again surface-active. That would explain both the second decrease in surface tension as well as the increase in surface elasticity. The phenomenon of the decrease in surface tension is also observed for other polyelectrolytes.[47]

4.3.4 Foam film stabilities.

As mentioned earlier the reduction of charge induces the destabilisation of the film. Comparing the stability curve on one hand with the surface tension and elasticity on the other hand one can easily see that they do not correlate with each other. Usually, a low surface tension or an elastic surface would increase the stability. Moreover, the stability of the films seems to depend only on the excess charge of the polyelectrolyte/surfactant system but not on the surface charge as there is no implication of a high polyelectrolyte surface coverage for the most stable film.

4.4 Conclusions

The results of the present chapter show that the reduction of the charge due to polyelectrolyte/surfactant complexation in the foam film leads to strong reduction in their stability. The minimum in stability of films containing surfactants and oppositely charged polyelectrolytes occurs around the nominal IEP of the system. Below the IEP the reduction in net surface charge is explained by the adsorption of polyelectrolytes at the surface and the formation of surface-active surfactant/polyelectrolytes complexes. This increase of material leads to a decrease in surface tension and an increase in surface elasticity. The increase in film stability above the IEP cannot be simply explained by a charge reversal at the interface due to an excess of polyelectrolytes. On the contrary, the increase in surface tension and the decrease in elasticity rather imply a loss in material. The fact that the stability and the surface potential are higher than far below the IEP indicates that the excess of polyelectrolytes within the film core contributes to the electrostatic repulsion rather than to a screening.

5 Effect of surfactant and polyelectrolyte hydrophobicity*

Abstract

The present study focuses on the stability control of foam films from oppositely charged polyelectrolyte/surfactant mixtures, namely, on the cationic surfactant dodecyl trimethyl ammonium bromide (C_{12}TAB) mixed with highly negatively charged polyelectrolytes. The excess charge of the polyelectrolyte/surfactant complexes can be tuned by varying the polyelectrolyte concentration so that foam films around the IEP can be studied. The measurements of the disjoining pressure isotherms show that the polyelectrolyte/surfactant ratio has an important influence on film stability, with a stability minimum close to the IEP and very stable films above that point. However, in the concentration regime in which the most stable films are formed, the surface coverage is very low, implying that the overall charge of the polyelectrolyte/surfactant complexes in the film dominates the film stability and not the complexes at the surface. Comparison with a previous work showed that the choice of surfactant plays an important role for tuning the foam stability while the type of polyelectrolyte has only a minor impact.

5.1 Introduction

Foams are dispersions of air in water and are widely used in industrial applications such as enhanced oil recovery and in personal care products. Depending on the field where it is applied, the foam should be either very stable or completely avoided. The stability of the foam depends essentially on the stability of the thin films that separate the dispersed phase, so that it is crucial to understand the properties of the single foam films. These films are stabilised by an excess pressure normal to the interfaces, the disjoining pressure. Depending on the origin of this force, two different film types can be distinguished: If electrostatic repulsion is the main force, a rather thick common black film (CBF) is formed. It has a thickness of about 100-10 nm and appears light or dark grey, depending on the respective film thickness. In case of a Newton black film (NBF), the film is mainly stabilised by steric forces and appears black due to its small thickness. This film type consists only of two layers of surfactant molecules, including their hydration shell, while all other liquid is pressed out of the film. That explains the thickness of about 5 nm, which corresponds to twice the length of a low molecular weight surfactant.[7]

The addition of polyelectrolytes strongly affects the properties of the foam films. Depending on the polyelectrolyte/surfactant mixture, different types of foam films can be observed: Likely charged components form CBFs due to the repulsion between the molecules at the interface and in the film core.[49,51,88] The same happens for oppositely charged mixtures in the semidilute concentration regime where surfactant/polyelectrolyte complexes are formed at the surface that

---

* Similar content has been published in : *Foam films from oppositely charged polyelectrolyte/surfactant mixtures: Effect of polyelectrolyte and surfactant hydrophobicity on film stability*, N. Kristen, A. Vüllings, A. Laschewski, R. Miller and R. v. Klitzing, *Langmuir*, **2010**, *26*, 9321-9327

can repel the polyelectrolytes in the core.[45–47] The only case where a NBF has been observed so far has been that of mixtures of nonionic surfactant and positively charged polyelectrolyte.[15,87] This is related to the fact that the air/water interface is slightly negatively charged[8] so that the polycation is attracted by the surface.

In the semidilute concentration regime, above a certain polyelectrolyte concentration, a stepwise thinning of the film thickness occurs. This so called stratification process arises from an oscillation of the disjoining pressure, that is originated from a transient network formed by the polyelectrolytes in the film core.[15,47,48,50,51,53,57] This network is concentration-dependent and is only formed when the critical overlap concentration, c^*, is reached. These results are in agreement with findings in bulk solutions[52,88] and can be generalised for many different polymers. When the pressure that is applied to the films is increased, layers of the network can be pressed out of the film. This phenomenon is independent of the surfactant type and can also be found between solid surfaces.[52]

In the present study, we focus on foam films from oppositely charged polyelectrolyte/surfactant mixtures. Above the critical surface aggregation concentration (csac), highly surface-active complexes are formed from the two components.[21,22] This occurs already at low polyelectrolyte concentrations, far below the concentration where aggregates are formed in the bulk (critical aggregation concentration, cac). These surface complexes have been the subject of many studies and have been characterised, for example, by means of surface tension,[25,27] elasticity measurements,[31,40,86] or various types of reflectivity methods.[24,26,28,39,84,85] In this work, we concentrate on mixtures from positively charged surfactants and negatively charged polyelectrolytes. Foam films with high polyelectrolyte concentrations are already well established, and the same applies to pure surfactant films. In both cases, CBFs are formed given that the surfactant can stabilise the film. We investigate the foam film properties between these two points, starting at very low polyelectrolyte concentrations up to concentrations where the amount of polymer exceeds that of the surfactant. In the beginning, the positive charge[89] of the surfactant dominates the film, while at high polyelectrolyte concentrations there is an excess of negative charges from the polymer. However, in between, there is a point, where the charge of one component is compensated by the other. In that case, at the IEP, where the repulsion is reduced, two scenarios are possible: the transition to a NBF or the destabilisation of the film.

In chapter 4, foam films from C_{14}TAB/PAMPS mixtures were studied. C_{14}TAB is the shortest surfactant in the C_nTAB series that forms stable films,[17] which makes it possible to investigate the whole concentration regime from films of pure surfactant solutions to those with a high amount of polymer. The stability of the foam films showed a clear dependence on the polyelectrolyte/surfactant ratio. At the IEP, it was not possible to form stable films at all, but once the IEP was crossed, the stability of the foam films increased again, even beyond that of the pure surfactant film. Surprisingly, the surface tension as well showed a breaking point just above the IEP, indicating a release the polyelectrolyte from the surface. Despite this release of surface complexes above the IEP, very stable films are formed in this concentration regime. This leads to the conclusion, that the overall charge in the system, including all charges of the surfactant and the polyelectrolyte at the surface and in the film bulk, is crucial for the stability and not only the surface charge, which has been the former assumption.

In the following, the effect of the hydrophilic/hydrophobic balance of the system on film stability is discussed. For this reason, the influence of another surfactant, namely C_{12}TAB, is investigated and the effect of polymer hydrophobicity is tested by exchanging PAMPS by PSS.

5.2 Results

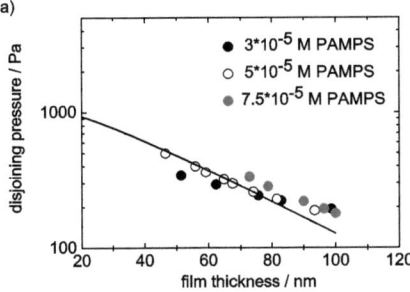 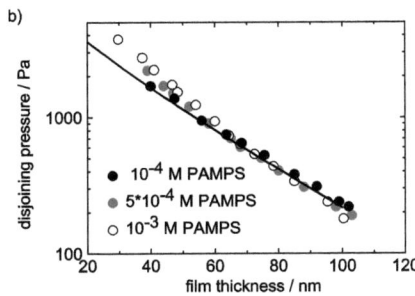

Figure 5.1: *Disjoining pressure isotherms of $C_{12}TAB/PAMPS$ solutions with a surfactant concentration of 10^{-4} M and different PAMPS concentrations: a) below the nominal IEP; b) at and above the nominal IEP. The solid line corresponds to the simulation of the isotherm with constant potential; for the sake of clarity, only some of the simulated isotherms are shown in the graph.*

5.2 Results

In the present chapter, we have investigated polyelectrolyte/surfactant mixtures of a fixed surfactant concentration and variable polymer concentration. It is important that the concentration regime is chosen in a way that it is well below the cmc, cac and c^*. Otherwise, these effects could interfere with the effect of the charge in the system, which is the focus of this study. The surfactant concentration was fixed at 10^{-4} M, which leads to a nominal IEP of 10^{-4} M in all experiments. The polyelectrolyte concentration is varied between 10^{-5} M and 10^{-3} M and refers to the concentration of monomer units (monoM).

5.2.1 C_{12}TAB/PAMPS mixtures

C_{12}TAB, unlike C_{14}TAB, cannot form stable films from pure surfactant solutions due to its shorter hydrophobic tail and the resulting lower Gibbs elasticity.[17] Only the addition of a certain amount of polyelectrolyte to the system allows the formation of stable films with C_{12}TAB which makes it impossible to study foam films at very low polymer concentrations. Fig. 5.1 shows disjoining pressure isotherms at a surfactant concentration of 10^{-4} M with a varied PAMPS concentration. The isotherms are divided into two groups: Fig. 5.1a shows isotherms below the nominal IEP, while Fig. 5.1b shows those at and above the nominal IEP. At first, the two groups seem to behave very differently, but a close look reveals the striking fact that all isotherms start at a equilibrium thickness of about 100 nm at low disjoining pressures, regardless of which amount of polyelectrolyte is added. This means that the isotherms coincide and differ only in slope and maximum pressure (Π_{max}). The first effect is related to the ionic strength in the system, since the slope increases with increasing ion concentration. The stability is defined as the maximum pressure that can be applied to the film before rupture.

In case of C_{12}TAB/PAMPS mixtures, a minimum concentration of at least 3×10^{-5} M PAMPS is needed to get a stable foam film with a maximum pressure of 300 Pa. An increase in polyelectrolyte concentration leads at first to an increase in stability to 500 Pa, but at 7.5×10^{-5} M PAMPS, close to the nominal IEP of 10^{-4} M, the stability decreases again to 300 Pa. At concentrations of 10^{-4} M or higher, very stable films are formed, with an increase in stability

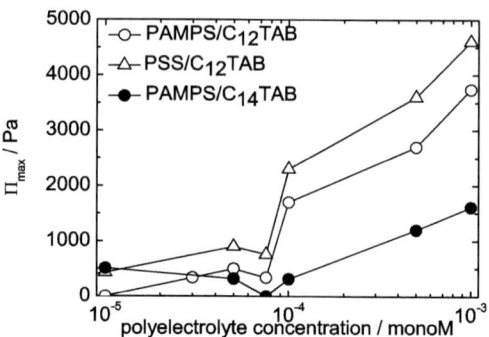

Figure 5.2: *Stability of the polyelectrolyte/surfactant films; maximum disjoining pressure Π_{max} before film rupture versus polyelectrolyte concentration at a fixed surfactant concentration of 10^{-4} M.*

from 1800 Pa at 10^{-4} M to 4000 Pa at 10^{-3} M PAMPS. These results are summarised in Fig. 5.2, which illustrates the film stabilities by plotting the maximum pressure Π_{max} versus the polyelectrolyte concentration. This stresses the enormous stability increase to 4000 Pa above the IEP compared to 500 Pa below and shows the stability minimum at 7.5×10^{-4} M.

To get more information about the surface potentials, simulations of the disjoining pressure isotherms of the respective mixtures were performed. The simulations were done at constant potential with the PB program version 2.2.1 by Per Linse et al.,[12] solving the nonlinear Poisson-Boltzmann equation. The results of the simulations are summarized in Tab. 5.1 and compared to the Debye length and the corresponding ionic strength calculated from the experimental data. The simulations of the disjoining pressure isotherms gave surface potentials between 52 and 62 mV below the IEP with a slight decrease at the stability minimum. The respective ionic strengths were close to 10^{-4} M, indicating that the surface potential in this concentration regime is dominated by the surfactant. The surface potentials of the foam films above the stability minimum are much higher, starting at 82 mV for 10^{-4} M PAMPS up to 92 mV at 10^{-3} M. The ionic strengths that were used for the simulation, were 1.2×10^{-4}, 2.0×10^{-4}, and 3.0×10^{-4} M. The values used for the simulations are in good agreement with the Manning concept of counterion condensation which predicts, that around every third polymer segment is dissociated.

To gain a deeper insight in the surface coverage and its correspondence to foam film stability, surface tension and elasticity measurements were performed. The surface tension measurements in Fig. 5.3 show a lowering of the surface tension below the IEP compared to the pure surfactant. At a PAMPS concentration of 10^{-5} M, the surface tension is already decreased to 66 mN/m and a minimum of 63 mN/m is reached at the nominal IEP. Further increase in polyelectrolyte concentration leads to a rise of the surface tension to a value close to that of the pure surfactant (about 70 mN/m), indicating a release of material from the surface. Only when the PAMPS concentration exceeds 10^{-3} M, the surface tension decreases again, an effect that has been observed for other polyelectrolytes in this concentration regime as well.[47]

Surface elasticity measurements (cf. Fig. 5.4) support the findings of the surface tension measurements. At PAMPS concentrations below 10^{-4} M, the elasticity increases from 10 mN/m

5.2 Results

PAMPS conc. [monoM]	Ψ_0 [mV]	κ^{-1} [nm]	I [M]
3×10^{-5}	52	30.7	9.0×10^{-5}
5×10^{-5}	62	29.7	1.0×10^{-4}
7.5×10^{-5}	60	29.0	1.1×10^{-4}
1×10^{-4}	82	25.1	1.2×10^{-4}
5×10^{-4}	87	20.7	2.0×10^{-4}
1×10^{-3}	92	17.7	3.0×10^{-4}

PSS conc. [monoM]	Ψ_0 [mV]	κ^{-1} [nm]	I [M]
1×10^{-5}	47	29.7	8.0×10^{-5}
5×10^{-5}	62	26.0	9.0×10^{-5}
7.5×10^{-5}	60	25.6	1.0×10^{-4}
1×10^{-4}	70	24.6	1.2×10^{-4}
5×10^{-4}	75	19.6	1.9×10^{-4}
1×10^{-3}	85	16.9	3.0×10^{-4}

Table 5.1: *Summary of the surface potentials Ψ_0 from the simulation of the disjoining pressure isotherms of polyelectrolyte/$C_{12}TAB$ films; the Debye length κ^{-1} is calculated from a fit of the experimental data with a exponential decay function of first order, the ionic strength I is derived from the simulation of the surface potential.*

(at 10^{-5} M) to 25 mN/m (at 5.0×10^{-5} M) and remains in this range until the nominal IEP, suggesting hydrophobic complexes at the surface. Once this point is crossed, the surface elasticity drops to a low value (5 mN/m), which corresponds to the rise in surface tension described above. After reaching a minimum, the surface elasticity rises again with increasing polyelectrolyte concentration.

5.2.2 C_{12}TAB/PSS mixtures

To study the effect of the hydrophobicity of the polyelectrolyte, a more hydrophobic polymer, namely PSS, has been chosen. This polyelectrolyte is linear and highly negatively charged and is more hydrophobic than PAMPS, due to the styrene unit in the backbone.

Disjoining pressure isotherms in Fig. 5.5 show that, in case of PSS, a lower amount of polyelectrolyte is needed to stabilise the films than in case of C_{12}TAB/PAMPS, since it is possible to form films already at a concentration of 10^{-5} M PSS. The film stability increases from 500 Pa to 1000 Pa at a concentration of 5.0×10^{-5} M and shows a minimum of 800 Pa at 7.5×10^{-5} M PSS (*cf.* Fig.5.2). At polyelectrolyte concentrations larger than 10^{-4} M, the foam films are significantly more stable, starting at 2300 Pa at 10^{-4} M, up to 5000 Pa at a concentration of 10^{-3} M PSS. The values of the surface potential describe well the shape of the film stability curve: Below the nominal IEP, surface potentials of 47, 62, and 60 mV were simulated for the films with a polyelectrolyte concentration below the nominal IEP. The assumed ionic strengths in these simulations are close to 10^{-4} M, the respective surfactant concentration in the measurements. With increasing PSS concentrations, the surface potentials increase as well, from 70 mV at 10^{-4} M to 85 mV at PSS concentration of 10^{-3} M. All disjoining pressure isotherms have an equilibrium thickness of 90 to 100 nm at pressures of about 200 Pa and get steeper with increasing polyelectrolyte concentration. This effect is more pronounced for isotherms at and above the nominal IEP and is related to the increasing ionic strength in the system. The ionic strengths used in the simulations, summarized in Table 5.1, were very close to the ones of the PAMPS/C_{12}TAB system and in good agreement with the Manning concept of polyelectrolyte

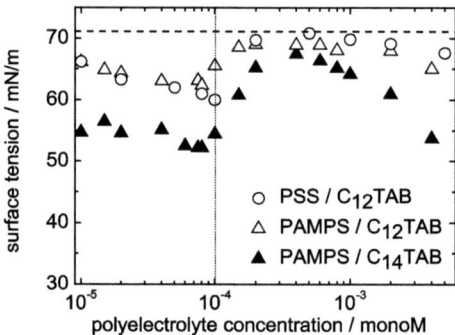

Figure 5.3: *Surface tension of different polyelectrolyte/surfactant solutions with fixed surfactant (10^{-4} M) and variable polyelectrolyte concentration; the dashed line corresponds to the surface tension of pure $C_{12}TAB$ and $C_{14}TAB$; the vertical line illustrates the nominal IEP of the system.*

condensation. In general, the shape of the stability curve of the foam films in dependency of the PSS concentration resembles that of PAMPS/C_{12}TAB but is shifted to higher stabilities.

The surface tension of the PSS/C_{12}TAB solutions shows basically the same characteristics as that of the other polyelectrolyte/surfactant mixtures. Below the nominal IEP, the surface tension is reduced compared to the pure surfactant solution, with a minimum of 60 mN/m at 10^{-4} M PSS. When the concentration of PSS is higher than 10^{-4} M, a sudden increase of the surface tension to 70 mN/m occurs and with further addition of the polyelectrolyte (more than 10^{-3} M) it decreases again. In the concentration regime below the IEP, the surface tension is slightly shifted to lower values compared to that of the PAMPS/C_{12}TAB system, whereas it shows slightly higher values after an increase of the surface tension. Generally, the release of surface complexes that is indicated by this rise is more distinct in the two systems with the surfactant with the shorter hydrophobic tail.

The measurements of the surface elasticity (Fig. 5.6) show a steady rise in elasticity, beginning at 15 mN/m at 10^{-5} M PSS and a maximum of 40-45 mN/m at 10^{-4} M, which is much higher than the maximum of PAMPS/C_{12}TAB. With increasing PSS concentration, the frequency dependence of the measurement increases as well. After the maximum is reached, the elasticity drops to the value of the pure surfactant and remains constant up to a concentration of 10^{-3} M, which is consistent with the surface tension measurements of the respective solutions.

5.2.3 C_{14}TAB/PSS mixtures

For the sake of completeness, C_{14}TAB/PSS mixtures have been investigated as well, but it has turned out that it is impossible to form stable and homogeneous films from these solutions, especially below the IEP. The surface tension measurements (*cf.* Fig. 5.7) show a very strong reduction in surface tension below the IEP with a minimum of 42 mN/m. This is the lowest value in all polyelectrolyte/surfactant systems that we have investigated and indicates a very strong interaction between the surfactant and the polyelectrolyte at the surface. Just above the nominal IEP, there is a sudden release of the hydrophobic complexes from the surface and the

5.3 Discussion

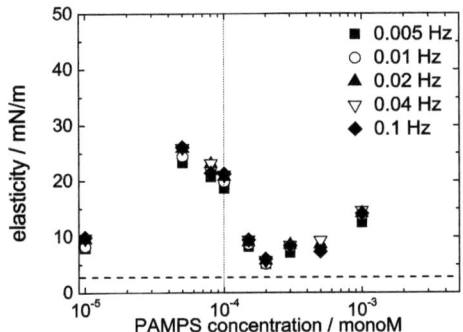

Figure 5.4: *Surface elasticity of $C_{12}TAB/PAMPS$ solutions with fixed surfactant (10^{-4} M) and variable PAMPS concentration at different frequencies; the dashed line corresponds to the elasticity of the pure $C_{12}TAB$ solution; the vertical line illustrates the nominal IEP of the system.*

surface tension rises to 70 mN/m again, which is very close to the value of the pure surfactant. This release is very sharp, like it is for the case of C_{12}TAB/PSS, and the addition of polymer up to a concentration of 2.0×10^{-3} M does not affect the surface tension significantly. Only at polyelectrolyte concentrations higher than this, the surface tension starts to decrease again.

The surface elasticity in Fig. 5.8 shows a non-monotonous behaviour. Even at a low polyelectrolyte concentration of 10^{-5} M, the surface is very elastic, with a huge frequency dependence (50-70 mN/m). With further addition of polyelectrolyte, the frequency dependence remains but the surface elasticity drops to 10-30 mN/m. At 10^{-4} M PSS, the nominal IEP, the surface elasticity is maximal, and in contrast to the lower concentrations the dependence on the frequency is lower. Above this point, the surface elasticity drops to a value that is almost not detectable anymore and remains constant in the investigated concentration regime.

5.3 Discussion

The focus of this study was the investigation of the properties of foam films from oppositely charged polyelectrolyte/surfactant mixtures around the IEP. In our former work, we studied foam films from aqueous PAMPS/C_{14}TAB mixtures.[90] The disjoining pressure isotherms showed a reduced stability with decreasing net charge of the polyelectrolyte/surfactant complexes and a complete destabilisation close to the nominal IEP. When this point was crossed, the stability of the foam films increased again (*cf.* Fig. 5.2) but the surface characterisation of the respective solutions revealed that this was not due to a charge reversal at the interface, which has been the working hypothesis so far. Surface tension and elasticity showed that there are indeed surface-active complexes at the interface below the IEP. However, when the polyelectrolyte concentration exceeds the surfactant concentration, which is the case at concentrations higher than 10^{-4} M, the surface tension increases and the elasticity drops, both close to the value that corresponds to that of the pure surfactant system. This indicates a release of the complexes from the surface, leaving a more or less pure surfactant layer at the surface. Only when the polyelectrolyte concentration is increased to the range where the stratification process

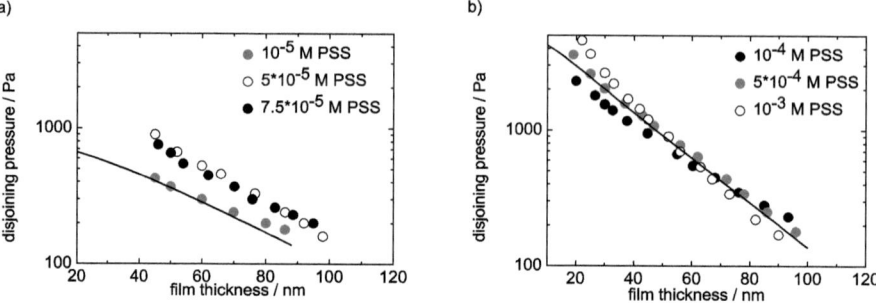

Figure 5.5: *Disjoining pressure isotherms of $C_{12}TAB/PSS$ solutions with a surfactant concentration of 10^{-4} M and different PSS concentrations: a) below the nominal IEP; b) at and above the nominal IEP; the solid line corresponds to the simulation of the isotherm with constant potential; for the sake of clarity, only some of the simulated isotherms are shown in the graph.*

starts, surface-active complexes occur again.

Surface tension and elasticity measurements of the surfactant/polyelectrolyte mixtures that are addressed in this study show the same non-monotonous behaviour. Below the nominal IEP, more surfactant molecules than polyelectrolyte segments are available. In this situation, the charges of the polymer units can be complexed by the surfactant, thereby exposing its hydrophobic tails to the air. This process is driven by the release of the counterions, which increases the entropy of the system[27] and leads to surface-active complexes that strongly reduce the surface tension. In the concentration regime above the IEP, the polyelectrolyte concentration exceeds that of the surfactant so that only part of the polymer segments can be decorated with surfactant molecules. This results in less hydrophobic complexes that are released from the surface, causing the increase of the surface tension. At higher polyelectrolyte concentrations, the semidilute concentration regime is reached and a network of polymers starts to form in the film core. In that case, the surfactant seems to be unequally distributed over this network of polymer chains so that some of them carry more surfactant molecules since they are close to surface. This would explain the reduction in surface tension and the increase in elasticity in this concentration regime.

5.3.1 Influence of the surfactant

The surfactant $C_{12}TAB$ has considerably different properties than $C_{14}TAB$. Although both surfactants have the same surface tension at the investigated concentration, foam films of the pure surfactant solutions show different properties: $C_{14}TAB$ forms fairly stable films, while $C_{12}TAB$ cannot stabilise foam films. It has a shorter hydrocarbon tail, which leads to a higher solubility in water as it is reflected in the cmc that is almost one order of magnitude higher than that of $C_{14}TAB$, and in lower adsorption rates to the surface.[91] A study by Bergeron[17] showed that the molecular area of a $C_{12}TAB$ molecule at the surface is significantly higher than that of $C_{14}TAB$ at the same surfactant concentration close to the cmc and can be explained by stronger van der Waals forces between the chains in the latter case. This leads to a more loosely packed $C_{12}TAB$ monolayer at the film surface compared to $C_{14}TAB$. This very dilute surface layer in turn results in a lower surface charge for $C_{12}TAB$ films which could explain the

5.3 Discussion

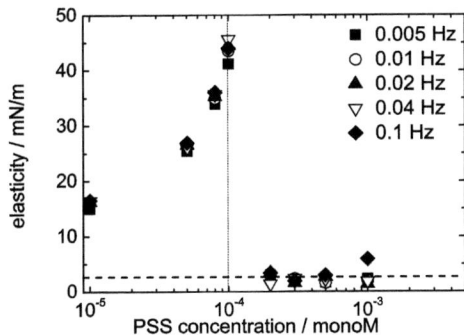

Figure 5.6: Surface elasticity of $C_{12}TAB/PSS$ solutions with fixed surfactant (10^{-4} M) and variable PSS concentration at different frequencies; the dashed line corresponds to the elasticity of the pure $C_{12}TAB$ solution; the vertical line illustrates the nominal IEP of the system.

difference in film stability between the two surfactants.

Comparing the shape of the stability graph of the different polyelectrolyte/surfactant systems, one clearly sees that the shape is mainly dependent on the surfactant type. In the case of $C_{14}TAB$, the addition of polyelectrolyte below the IEP leads to the reduction of the stability, whereas in foam films with $C_{12}TAB$ the films first get more stable when polymer is added. This increase in stability is accompanied by an increase in surface potential (from 52 to 62 mV in case of PAMPS and from 47 to 62 mV for PSS), indicating that the polymer contributes to bring material to the surface by screening the charge of the surfactant, rather than just reducing the charge at the interface.

The second important difference of the foam film stabilities is the minimum in stability that is observed close to the IEP. Stable films can be formed at the IEP when $C_{12}TAB$ is used as a surfactant, whereas in case of $C_{14}TAB$ the film was completely destabilised. Fig. 5.3 reveals that the reduction in surface tension in this regime is much less pronounced for $C_{12}TAB$ than it is for $C_{14}TAB$. This indicates that the complexes that are formed between the surfactant and the polymer are less hydrophobic and therefore less surface-active, which suggests that the transport of the material to the surface is reduced. This could lead to unbound molecules at the surface giving rise to the surface potential of 60 mV at the point of minimum stability. This value is slightly lower than that of the foam film with lower polyelectrolyte concentration due to the fact that more polyelectrolyte/surfactant complexes are formed.

The fact that the observed IEP is slightly below the point, where the ratio of polyelectrolyte and surfactant is 1, could have various reasons: for example, the composition at the surface could be different from that in the bulk solution, or the dissociation of the surfactant could be incomplete. However, this point of minimum stability was found to be at a polyelectrolyte concentration 7.5×10^{-5} M in all studied polyelectrolyte/surfactant systems.

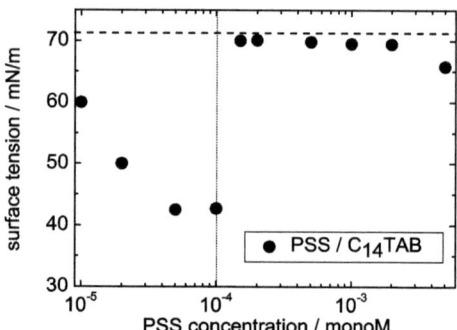

Figure 5.7: *Surface tension of $C_{14}TAB/PSS$ solutions with fixed surfactant (10^{-4} M) and variable PSS concentration; the dashed line corresponds to the surface tension of the pure $C_{14}TAB$ solution; the vertical line illustrates the nominal IEP of the system.*

5.3.2 Influence of the polyelectrolyte

Exchanging the polyelectrolyte has only a small impact on the properties of foam films with $C_{12}TAB$. Unfortunately, in this work, only changes of film stability in systems with this surfactant can be addressed, since it is impossible to form stable and homogeneous films from PSS/$C_{14}TAB$ solutions.

When the two systems with different polyelectrolytes are compared, one clearly sees that the film stabilities follow the same trend but that the PSS films are more stable. However, the simulated potentials of the film surfaces are in the same range for both polyelectrolytes, implying that this is not due to electrostatic reasons but rather has another mechanism. PSS, unlike PAMPS, can expose its hydrophobic parts to the surface, which can have an additional stabilising effect on the films. Below the IEP, this assumption is supported by a slightly lower surface tension and the continuous increase of the surface elasticity upon addition of polyelectrolyte. This additional stabilising element at the surface would also explain why a lower PSS concentration is needed to form stable films.

However, above the IEP, this simple picture cannot be sustained anymore. In this concentration regime, most of the polyelectrolyte/surfactant complexes are released from the surface, and surface tension and elasticity measurements do not give rise to the assumption that the polyelectrolyte contributes to the stabilisation of the foam films by adsorbing to the interface. Quite to the contrary, the release of surface complexes is even more pronounced for PSS, which is expressed by a sharp jump in surface tension and surface elasticity. The jump is followed by a rather long plateau, where both the surface tension and elasticity remain constant, which is observed for mixtures with both surfactants. A possible explanation for this effect could be the fact that the hydrophobicity of PSS can give rise to hydrophobic interactions between the polyelectrolyte and the surfactant.[33] The hydrophobic tail of the surfactant can interact with the hydrophobic parts in the core of the polyelectrolyte coil, thereby exposing the charged headgroup to the outside and making the complex even more hydrophilic. The effect is probably more pronounced for PSS/$C_{14}TAB$. These hydrophilic complexes could be a reason for the increased stability of the foam films from PSS/$C_{12}TAB$ mixtures, since additional charges

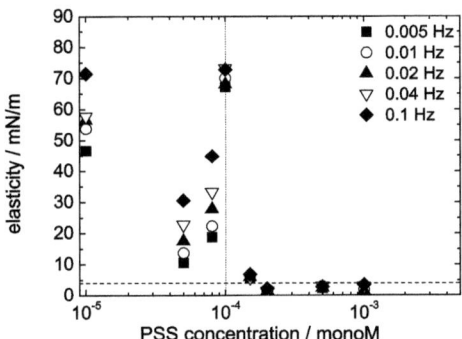

Figure 5.8: *Surface elasticity of $C_{14}TAB/PSS$ solutions with fixed surfactant (10^{-4} M) and variable PSS concentration at different frequencies; the dashed line corresponds to the elasticity of the pure $C_{14}TAB$ solution; the vertical line illustrates the nominal IEP of the system.*

are exposed to the film bulk. Yet, the effect is not reflected in the surface potentials, so it remains rather speculative. In general, the measurements show that the polyelectrolytes in the film bulk have a strong influence on the film stability, even when they are not located at the surface. Very stable films are formed in a concentration regime with very low surface coverage; moreover, the most stable films are observed for PSS/C_{12}TAB where the surface tension is that of the pure surfactant. This implies that the properties of the foam are influenced not only by the surface properties but also by the bulk composition.

As mentioned above, PSS/C_{14}TAB is a special case among the different polyelectrolyte/ surfactant mixtures that are investigated in this study. The low surface tension below the IEP suggests a very high surface coverage, but no stable film can be formed in this concentration regime. The reduction in surface elasticity below the IEP could be a hint for a very rigid surface coverage. The attraction between the two compounds is very strong, since both hydrophobic and electrostatic interactions are possible. Due to these strong interactions, a network of polyelectrolyte/surfactant complexes can be formed at the surface, which is cross-linked by the C_{14}TAB molecules. This network leads to a reduction in surface elasticity, and therefore, to unstable films. Additional problems arise when the PSS concentration is further increased. In the concentration regime at and above the IEP, larger aggregates are formed in the bulk solution. When a film is formed from these solutions, the aggregates are trapped in the film core, leading to inhomogeneous films, that cannot be investigated.[22] Altogether, this makes it impossible to include this system in the reflections on film stabilities.

5.4 Conclusions

The present study shows that the surfactant plays a major role in the stabilisation of foam films. In presence of C_{12}TAB, a certain amount of polyelectrolyte is needed to get stable films. The addition of polymer leads to stable films throughout the whole concentration regime and only a minimum stability is observed at the IEP point. Above the IEP, very stable films are formed, even though the surface characterisation methods indicate only a small amount of material at the surface. In this work, the most stable films occurred in the polyelectrolyte/surfactant system with the highest surface tension, implying the lowest surface coverage at high polyelectrolyte concentrations. This confirms the hypothesis that not only the polyelectrolyte at the surface but also that in the film core contribute to the stabilisation. The variation of the polyelectrolyte had only a minor effect on the film stability; the exchange of the polymer to the more hydrophobic PSS resulted in a shift to slightly higher maximum pressures, which can be explained by the integration of the hydrophobic parts of the polyelectrolyte within the surface.

6 Variation of the isoelectric point

Abstract

In this chapter, the influence of the IEP on the film properties was studied and compared to the findings of PAMPS/C_{14}TAB with an IEP of 10^{-4} M. To verify the assumption that the position of the IEP does not change the typical character of the foam film stabilities, the IEP of the polyelectrolyte/surfactant mixtures has been shifted in two different ways. Within the first series of measurements, the IEP was changed by reducing the fixed surfactant concentration in the mixture, leading to a more diluted system. In the second approach, a copolymer of nonionic and ionic monomer units was used to lower the charge density of the polymer. This gave rise to the additional interactions between the polyelectrolyte and the surfactant, which makes the description of the foam film properties more complex. In both systems, the same characteristics of the foam film stabilities were found, but the concentration range, where unstable film were formed, was broader. However, the mechanisms leading to the destabilisation were different.

6.1 Introduction

Interactions between polyelectrolytes and surfactants are of great relevance for many applications. Depending on the type of polyelectrolyte and surfactant, the interactions can be of different nature. In case of charged polyelectrolyte/surfactant mixtures, either repulsive or attractive electrostatic interactions can occur between the components.[92] Furthermore, when nonionic or more hydrophobic compounds are used in a mixture, hydrophobic interactions play an important role as well.[93,94] All these forces have a strong influence on, *e.g.*, the properties of foam films or the hydrodynamic behaviour of the bulk solution.

In case of oppositely charged polyelectrolyte/surfactant mixtures, a strong electrostatic attraction is observed. The charges of the polyelectrolyte are complexed by surfactant molecules of opposite charge, which is both energetically and entropically favourable due to the release of counterions.[27] Depending on the ratio of the two compounds, the formation of these complexes can lead to hydrophobic aggregates located at the surface. Furthermore, the study of different polyelectrolyte/surfactant mixtures showed that the stability of foam films strongly depends on this ratio.

In this context, the stability of foam films around the IEP of the system was investigated (see chapter 4 and 5). To get a deeper insight into the properties of foam film around the IEP, foam films at different polyelectrolyte/surfactant ratios were studied. The investigation shows, that the stability of the foam films decreases when the IEP of the system is approached, and when the IEP is crossed, foam films with a high stability are formed. In case of PAMPS/C_{14}TAB, for example, a complete destabilisation takes place very close to the nominal IEP. Surprisingly, the surface coverage gives no hint for a charge reversal at the interface, which has been assumed to be the reason for the decreases of foam film stability around the IEP.

To verify the hypothesis that these findings are not unique for systems with an IEP of 10^{-4} M, it is important to investigate polyelectrolyte/surfactant mixtures where the position of the

IEP is shifted. For PAMPS/C_{14}TAB systems, two different ways of changing the point of equal charges are possible. In the first series of experiments, the concentration of the surfactant is changed, leading to a system that is either more concentrated or more dilute as the system described in chapter 4. Accordingly, only concentration effects are assumed to influence the interactions in foam films and hence, their stability.

An alternative to the variation of the surfactant concentration is the use of a polyelectrolyte with different degree of charge. As only polymers with a charge density of 100 % were studied in former investigations, a polymer with a lower degree of charge has to be used for this approach. For this purpose, a random copolymer of negatively charged and nonionic monomer units, namely poly[tris (hydroxymethy)methyl]acrylamide co acrylamido methyl propanesulfonate, is synthesised. Within this copolymer, the charged units are exchanged by nonionic but not very hydrophobic monomer units. Beside electrostatic attraction between the charged components, now also hydrophobic interactions between the nonionic monomer units and the surfactant or other polyelectrolyte chains can occur. In general, nonionic polymers interact stronger with negatively charged surfactants like sodium docecyl sulfate (SDS) and show only weak interactions with cationic surfactants. This is interpreted as a more favourable interaction of anionic surfactants with the hydration shell of the polymer or the unfavourable bulkiness of the cationic head group.[95] Nevertheless, interactions between nonionic polyelectrolytes and C_nTAB have been reported in literature,[93,95,96] and are supposed to be stronger for more hydrophobic polymers. Altogether, this implies that the interactions in case of C_{14}TAB mixtures with PAMPS with a lower degree of charge are more complex.

In the following, the results of the described investigations are presented and compared to the former findings.

6.2 Results

In a first set of experiments, the IEP was shifted to verify the hitherto assumption that, regardless of the surfactant concentration, PAMPS/C_{14}TAB solutions are destabilised close to the nominal IEP. An other surfactant concentration than 10^{-4} M, which shifts the nominal IEP, has to be carefully chosen. To get a significantly different IEP, the surfactant concentration should not be too close to 10^{-4} M. On one hand, when the amount of surfactant is too high, the corresponding polyelectrolyte concentration crosses c*, and the effect of the surface charge could be superposed by the stratification process, which would cover surface effects. On the other hand, a certain surfactant concentration is needed to stabilize the foam film, as this is a precondition for this type of investigation. After some preliminary tests, the C_{14}TAB concentration was fixed at 3×10^{-5} M, which fulfills the above described requirements of the experiment. As in previous chapters, the surfactant concentration is fixed in all experiments and only the polyelectrolyte concentration is varied.

In Fig. 6.1, disjoining pressure isotherms of PAMPS/C_{14}TAB films at polyelectrolyte concentrations between 10^{-6} M and 10^{-3} M PAMPS and of the foam film of the pure surfactant solution at 3×10^{-5} M are shown. In Fig. 6.1a, isotherms below the nominal IEP are depicted. Note, that the films are thicker now compared to all other disjoining pressure isotherms of polyelectrolyte/surfactant mixtures in this thesis. The foam film stabilised by the C_{14}TAB solution without any additional PAMPS is formed at an equilibrium thickness of 153 nm, which is 50 nm thicker than the corresponding foam film at 10^{-4} M C_{14}TAB. The addition of 10^{-6} M PAMPS induces screening of the surface charge and reduces the thickness at 200 Pa by 10 nm to 142 nm. Disjoining pressure isotherms of foam films with a PAMPS concentration larger than 10^{-4} M are shown in Fig. 6.1b. In this case, the further screening in the films leads to an

6.2 Results

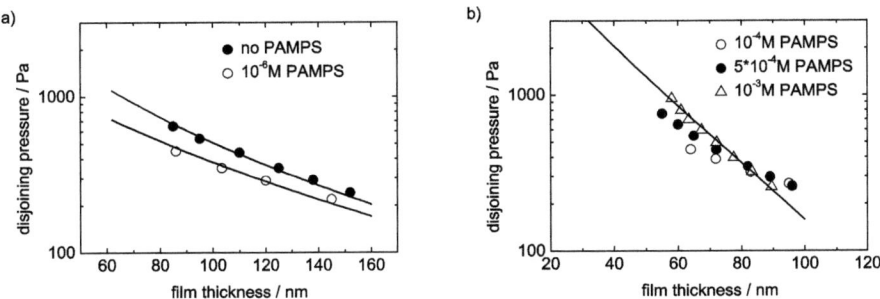

Figure 6.1: *Disjoining pressure isotherms of PAMPS/C_{14}TAB solutions with a fixed surfactant concentration of 3×10^{-5} M and with varied polyelectrolyte concentration; a) below the IEP; b) above the IEP; the solid lines represent the simulations of the foam films at constant potential; for the sake of clarity, only some of the simulated isotherms are shown in the graph.*

equilibrium thickness between 90-100 nm at low disjoining pressures, which is in good agreement with the measurements at higher surfactant concentration. The slope of the respective isotherms increases with increasing polyelectrolyte concentration. This phenomenon is related to the fact that a higher polyelectrolyte concentration leads to a higher ionic strength in the system. Furthermore, the ionic strengths found in the mixtures (*cf.* Table 6.1)are in good agreement with the Manning concept of counterion condensation, which predicts for PAMPS that only two thirds of the polyelectrolyte charges are compensated by counter ions.

In Fig. 6.2 the stabilities of the all described foam films are depicted by plotting the maximum disjoining pressure before film rupture versus the polymer concentration. The pure C_{14}TAB foam film is stable up to a disjoining pressure of 700 Pa. The reduction of film stability compared to the film at a C_{14}TAB concentration of 10^{-4} M (Π_{max}= 900 Pa) can be explained by lower coverage with surfactant molecules at the surface at this concentration. With the addition of 10^{-6} M PAMPS, Π_{max} of the foam film decreases to 480 Pa. These findings are supported by the corresponding potentials of the film surface that are derived from simulations of the disjoining pressure isotherms. The program was written by Per Linse[12] and it solves the nonlinear Poisson-Boltzmann equation by assuming constant potential. Foam films from the pure surfactant solutions have a surface potential of 117 mV, which is very high compared to 78 mV of the film at 10^{-4} M C_{14}TAB. The potential is reduced to 102 mV, when 10^{-6} M PAMPS is added. In a concentration regime between 10^{-6} M and 10^{-4} M PAMPS no stable film can be formed at all. This is a broad concentration regime compared to the system of PAMPS/C_{14}TAB with an IEP at 10^{-4} M, where only one particular polyelectrolyte concentration leads to the destabilisation of the foam film. The addition of more polyelectrolyte leads to the increase in film stability from 450 Pa at 10^{-4} M to almost 1000 Pa at a PAMPS concentration of 10^{-3} M. This is accompanied by an increase in surface potential up to 90 mV at 10^{-3} M PAMPS. All results of the film simulations are summarised in Table 6.1.

The high surface potential and film thickness of the foam film at 3×10^{-5} M C_{14}TAB gives reason to a deeper investigation of the pure surfactant films. Fig. 6.3 shows isotherms of pure C_{14}TAB films at different surfactant concentrations. It reveals that the foam films get thinner and the isotherms steeper with increasing surfactant concentration. As mentioned earlier, both effects are related to the increasing ionic strength in the investigated surfactant solutions. As

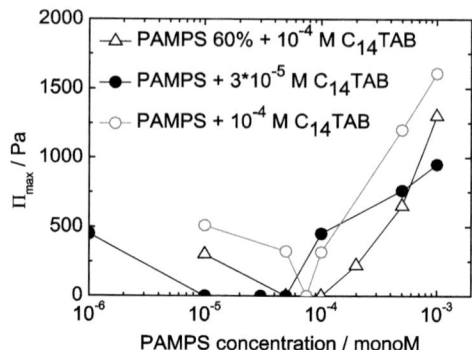

Figure 6.2: *Stability of PAMPS/C_{14}TAB films at different IEP; IEP = 3×10^{-5} M (filled circles); IEP = 1.6×10^{-4} M (empty triangles); IEP = 10^{-4} M (empty circles); maximum disjoining pressure Π_{max} before film rupture versus polyelectrolyte concentration.*

PAMPS conc. [monoM]	Ψ_0 [mV]	κ^{-1} [nm]	I [M]
no PAMPS	117	53.1	2.0×10^{-5}
1×10^{-6}	102	50.0	2.0×10^{-5}
1×10^{-4}	70	33.8	5.0×10^{-5}
5×10^{-4}	75	23.3	1.0×10^{-4}
1×10^{-3}	90	16.5	2.5×10^{-4}

Table 6.1: *Summary of the surface potentials Ψ_0 from the simulation of the disjoining pressure isotherms of PAMPS/C_{14}TAB films at a surfactant concentration of 3×10^{-5} M; the Debye length κ^{-1} is calculated from a fit of the experimental data with a exponential decay function of first order, the ionic strength I is derived from the simulation of the surface potential.*

described above, at the lowest investigated concentration, a thickness of 153 nm is observed. With the addition of higher surfactant concentrations, the film thickness considerably decreases from 101 nm at 10^{-4} M to 70 nm at 10^{-3} M due to the screening of the surfactant head groups at the surface. All film thicknesses mentioned in this paragraph refer to the first data point of the isotherm at around 220 Pa. In addition to the film thinning, the stability of the foam films steadily increases. At the lowest investigated C_{14}TAB concentration, Π_{max} is 700 Pa. With the addition of 10^{-4} M C_{14}TAB, the film stability increases to 900 Pa. This difference in stability is not very pronounced compared to the increase that is observed upon addition of 5×10^{-4} M C_{14}TAB. At this concentration, the film stability is with 4000 Pa significantly more stable. The trend continues to the foam film with the surfactant concentration of 10^{-3} M, where the films are stable to a disjoining pressure of 8000 Pa, close to the limit of the method. However, in contrast to the monotonous behaviour of the disjoining pressure isotherms in terms of film thinning and stability increase, the surface potentials obtained from the simulations of the isotherms are not so uniform. In fact, the potential is strongly reduced from 117 mV to 78 mV when the surfactant concentration is increased from 3×10^{-5} M to 10^{-4} M. At this concentration, a minimum potential is observed and the potential rises again, when more

6.2 Results

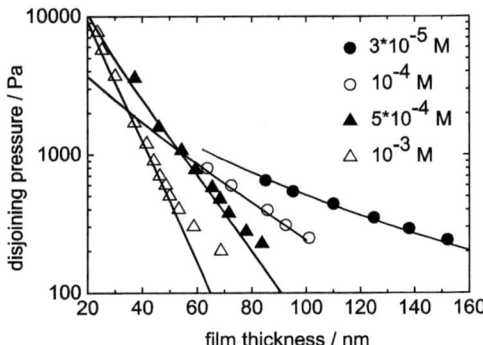

Figure 6.3: *Disjoining pressure isotherms of $C_{14}TAB$ at different concentrations; the solid lines correspond to the simulations at constant potential.*

C_{14}TAB conc. [M]	Ψ_0 [mV]	κ^{-1} [nm]	I [M]
3×10^{-5}	117	53.1	2.0×10^{-5}
1×10^{-4}	78	29.3	1.0×10^{-4}
5×10^{-4}	95	15.9	4.0×10^{-4}
1×10^{-3}	98	10.3	1.0×10^{-3}

Table 6.2: *Summary of the surface potentials Ψ_0 from the simulation of the disjoining pressure isotherms of $C_{14}TAB$ films at different surfactant concentrations; the Debye length κ^{-1} is calculated from a fit of the experimental data with a exponential decay function of first order, the ionic strength I is derived from the simulation of the surface potential.*

C_{14}TAB is added to the system. At 10^{-3} M, the surface potential reaches a value of almost 100 mV (see Table 6.2). For a better characterisation of the surface properties of foam film stabilised by PAMPS/ C_{14}TAB mixtures at low IEP, surface tension measurements were performed. Since it is not possible to measure the interfacial tension of the foam film that of the corresponding polyelectrolyte/surfactant solution has been investigated. In fact, theoretical considerations predict a different surface coverage in case of a foam film compared to the surface of the bulk solution,[73,97] but it is assumed that these changes are only minor so that they can be neglected. As shown in Fig 6.4, the shape to the surface tension isotherm of PAMPS/C_{14}TAB at 3×10^{-5} M resembles qualitatively that of the mixture at 10^{-4} M C_{14}TAB, only the IEP is shifted to a lower concentration and the reduction in surface tension is not so strong. In both cases, the addition of a low amount PAMPS (10^{-6} M) has almost no influence on the surface tension, so that it corresponds to that of the pure surfactant solution. When the polyelectrolyte concentration is increased in the systems with 3×10^{-5} M C_{14}TAB, the surface tension decreases steadily until a minimum of 65 mN/m, which is less pronounced than in case of a higher surfactant concentration. For PAMPS concentrations above the nominal IEP, the surface tension goes back to the value of the pure surfactant solution, which implies the release of material from the surface. However, polyelectrolyte concentrations larger than 2×10^{-4} M induce again the lowering of the surface tension.

Another way to shift the IEP of the system is to change the degree of charge of the polyelec-

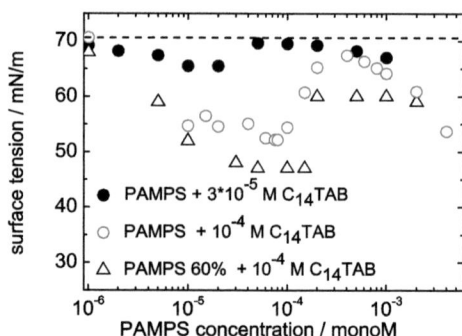

Figure 6.4: *Surface tension of PAMPS/C_{14}TAB solutions with fixed surfactant and variable PAMPS concentration at different IEP; IEP = 3×10^{-5} M (filled circles); IEP = 10^{-4} M (empty circles); IEP = 1.6×10^{-4} M (empty triangles); the dashed line corresponds to the surface tension of the pure surfactant at both concentrations.*

trolyte. So far, only polyelectrolytes with 100 % charged monomer units have been investigated. Again the type of polymer and the degree of charge need to be carefully chosen. On one hand, the degree of charge of the polyelectrolyte should be significantly different from 100 % to get a detectable shift of the IEP. On the other hand, the reduction of charge density increases the hydrophobicity of the polymer, which can lead to the formation of aggregates and to a decrease in water solubility. This phenomenon is for example observed for PSS, where the reduction of sulphonated monomers leads to surface-active polymers,[98] that are accessible to hydrophobic interactions with the surfactant (100 % charged PSS is not surface-active in the investigated concentration regime). However, in case of PAMPS, at a degree of charge of 60 %, no problems with solubility occur and the formation of aggregates is not very pronounced. The nominal IEP in this system is at 1.6×10^{-4} M.

In Fig. 6.5, the disjoining pressure isotherms of the PAMPS 60%/C_{14}TAB system are shown. In Fig. 6.5a, isotherms without any polyelectrolyte and with a PAMPS 60% concentration below the IEP are plotted, while in Fig 6.5b isotherms above the nominal IEP are shown. The foam film of the pure surfactant has a starting thickness of 101 nm. The addition of 10^{-5} M PAMPS 60% induces a slight increase in film thickness, which is very unusual. At higher polymer concentration, the equilibrium film thickness at 200 and 250 Pa, respectively, is around 100 nm which is rather high compared to other polyelectrolyte/surfactant systems. Furthermore, the slope of the isotherms increases with increasing ionic strength. The stability of the foam films is shown in Fig. 6.2. As described in detail in chapter 4, the pure surfactant foam film at 10^{-4} M is stable up to a disjoining pressure of 900 Pa. The addition of PAMPS 60% leads to a strong reduction in film stability up to a maximum pressure of 300 Pa. Between 10^{-5} M and 2×10^{-4} M PAMPS 60%, which is slightly above the nominal IEP of the system, no stable films can be formed. When the polyelectrolyte concentration is increased to 2×10^{-4} M, the corresponding foam film reaches a stability of 250 Pa, which is depicted in Fig 6.5b by a filled circle. Further increase of the PAMPS 60% concentration results in a strong increase in stability to 650 Pa at 5×10^{-4} M and 1200 Pa at 10^{-3} M, respectively. The shape of the stability curve is supported by the surface potentials that are simulated as specified above. The surface potentials of the foam films at 10^{-4} M and 2×10^{-4} M are not available, since there are too few data points

6.2 Results

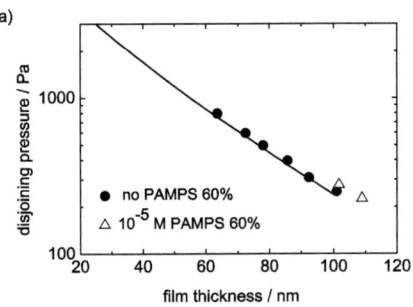
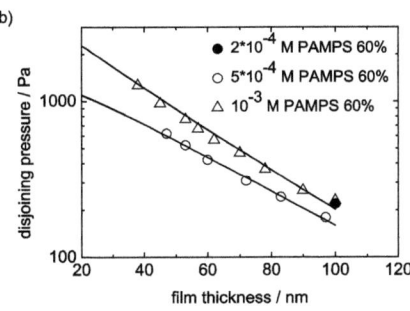

Figure 6.5: *Disjoining pressure isotherms of PAMPS 60%/C_{14}TAB solutions with fixed surfactant concentration (10^{-4} M) and varied polyelectrolyte concentration; a) below the IEP; b) above the IEP; the solid lines represent the simulations of the foam films at constant potential; for the sake of clarity, only some of the simulated isotherms are shown in the graph.*

to simulate the isotherm. The simulation of the isotherm without polyelectrolyte results in a surface potential of 78 mV. The next isotherm that is accessible to the simulation is the one at 5×10^{-4} M. This foam film has a potential of 58 mV, considerably lower than that of the pure surfactant film. With increasing polyelectrolyte concentration, the surface potential rises as well and reaches a value of 72 mV at 10^{-3} M PAMPS 60%.

PAMPS 60% conc. [monoM]	Ψ_0 [mV]	κ^{-1} [nm]	I [M]
no PAMPS	78	29.8	1.0×10^{-4}
5×10^{-4}	58	28.0	1.0×10^{-4}
1×10^{-3}	72	25.7	1.3×10^{-4}

Table 6.3: *Summary of the surface potentials Ψ_0 from the simulation of the disjoining pressure isotherms of PAMPS 60%/C_{14}TAB films; the Debye length κ^{-1} is calculated from a fit of the experimental data with a exponential decay function of first order, the ionic strength I is derived from the simulation of the surface potential.*

The shape of the surface tension isotherm of PAMPS 60%/C_{14}TAB in Fig 6.4 is slightly different compared to PAMPS/C_{14}TAB mixtures with a degree of charge of 100 %. The addition of polymer to the surfactant solution has a much larger influence on the surface tension even though the degree of charge and the corresponding ability to form complexes with the surfactant is lower. At 10^{-6} M PAMPS 60%, the surface tension is already reduced from 70 to 68 mN/m compared to the pure surfactant solution. Further increase of polymer concentration induces a strong reduction of surface tension until a plateau of 46 mN/m is reached at 5×10^{-5} M PAMPS 60%. The surface tension remains constant until the nominal IEP of 1.6×10^{-4} M. The plateau is 6 mN/m below the minimum that is obtained for PAMPS/C_{14}TAB with an IEP of 10^{-4} M. At higher polymer concentrations a sudden increase of surface tension can be observed, but in contrast to other polyelectrolyte/surfactant systems, the mixture does not reach the value of the pure surfactant. In fact, a second plateau is reached at 60 mN/m.

Since the surface tension measurements reveal rather unexpected characteristics of PAMPS 60%/C_{14}TAB complexes at the surface, additional experiments have been carried out. In Fig. 6.6 the results of dilational surface elasticity measurements are shown. These experiments were

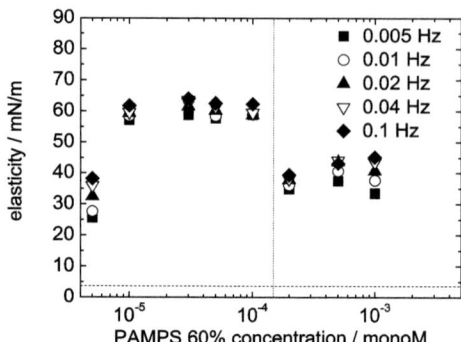

Figure 6.6: *Surface elasticity of PAMPS 60%/C_{14}TAB solutions with fixed surfactant (10^{-4} M) and variable PAMPS concentration at different frequencies; the dashed line corresponds to the elasticity of the pure surfactant; the vertical line illustrates the nominal IEP of the system.*

performed by using the oscillating drop method at frequencies between 0.005 and 0.1 Hz. As suggested by the surface tension measurements, the surface elasticity increases upon polyelectrolyte addition compared to the value of the pure surfactant, until a plateau is reached at 10^{-5} M PAMPS 60%. At the lowest measured polyelectrolyte concentration, the elasticity of the different frequencies varies between 25 and 40 mN/m. At the plateau of 60 ± 5 mN/m, the variation is smaller and is even more reduced close to the nominal IEP. Above the IEP, the surface elasticity drops to 38 ± 3 mN/m and stays constant up to a polymer concentration of 10^{-3} M, which is consistent with the surface tension measurements. In comparison to PAMPS/C_{14}TAB the elasticity shows a stronger frequency variation. Additionally, the variation of the surface elasticity due to the investigated different frequencies increases continuously in this concentration regime. Altogether, the elasticity measurements support the results of the surface tension measurements.

6.3 Discussion

The foam film stabilities of the two systems reveal similarities to the mixture PAMPS/ C_{14}TAB described in chapter 4. The addition of small amounts of polyelectrolyte reduces the film stability, compared to the film stabilised only by surfactant, until a total destabilisation occurs in a concentration regime close to the respective IEP of each system. When the concentration of polyelectrolyte segments exceeds that of the surfactant, the films are significantly more stable than the pure surfactant films. The mixtures described in this study show a broader concentration range where unstable films occur than PAMPS/C_{14}TAB. In addition, surface tension and elasticity measurements of PAMPS 60% show deviations from the findings of PAMPS/C_{14}TAB, so that the destabilisation is presumably based on different mechanisms in the two cases. These differences will be discussed in the following.

6.3 Discussion

6.3.1 Effect of the surfactant concentration

The major difference between the PAMPS/C_{14}TAB system with an IEP at 10^{-4} M and the one at 3×10^{-5} M, respectively, is the surfactant coverage at the interface. However, since at both surfactant concentrations the surface tension is close to water no difference in surface coverage is detectable.[16] In contrast to that, the foam films of the two investigated surfactant concentrations show significantly different interactions between the film surfaces. At 3×10^{-5} M, the disjoining pressure isotherms of the foam film are flatter, which can be explained by the low ionic strength of the system. Furthermore, the films are very thick. In this concentration regime, two effects counteract each other. On one hand, the electrostatic repulsion arising from the dissociated surfactant molecules is low. Yet, on the other hand, the screening of the electrostatic double layer is not very pronounced as well. However, both the film thickness and the surface potential are very high at at 3×10^{-5} M C_{14}TAB. Roughly spoken, the film thickness corresponds to 3× the Debye length of the system. The film thickness is assumed to be based on the high potential at the interface. The surface potential, in turn, results from an interplay of several contributions, like surfactant adsorptions at the interface, degree of dissociation of the charged molecule, charge screening and molecular conformation at the surface. Since experimental evidence gave rise to the assumption that the pure water/air interface is negatively charged,[9] the addition of cationic surfactant leads to a charge reversal at the interface. One major drawback of the surface potential simulations of symmetric films is that no statement can be made on the sign of the potential, so that other methods have to be taken into account to get information about it. For example, a study on wetting films of C_{14}TAB solutions reported the IEP of the surface to be at 10^{-6} M[99] which is well below the concentrations investigated in this work and leads to the conclusion that all surface potentials of the C_{14}TAB foam films are positive. This information reveals that there is indeed a minimum potential at a surfactant concentration of 10^{-4} M.

There are only few studies of the surface potential of pure surfactant foam films and to our knowledge, this phenomenon has not been observed for C_nTAB before. However, the study by Exerowa et al.[100] only shows the surface potential of foam films with a salt concentration of 5×10^{-4} M NaCl, which strongly influences the potential (see chapter 7 and Ref. 101), so that the two sets of results can not be directly compared. In contrast to that, a surface potential minimum has been found for SDS at the water/air interface as well.[101] This is explained by a change of conformation of the adsorbed molecules at the transition from a very dilute surface coverage to one that is more dense. At this point, the conformation of the surfactant molecules changes from a more or less flat adsorption to an upright position and the dipole moment of the molecule starts to contribute to the potential. This positive contribution of the dipole moment leads to a reduction of the surface potential to almost 0 mV at concentrations over 10^{-3} M. However, this theory can not be applied to the current system as it does not explain the high positive potential at low surface coverage.

Another possible explanation is given by Teppner et al.,[102] who found in an ellipsometry study that the counterion distribution changes near the interface depending on the investigated surfactant concentration. At low surfactant concentrations, the counterions do not enter the adsorption layer but stay in the diffuse double-layer near the interface, so that no ion pairs can be formed and no reduction of the surface charge occurs. At higher concentrations, the counterions are situated in both the adsorption and the diffuse double layer, so that charge screening starts to play a role. This would explain the surface potential of 117 mV at 3×10^{-5} M C_{14}TAB, with the absence of ion pair formation at the interface. When the surfactant concentration is increased to 10^{-4} M, the surface potential is reduced to 78 mV due to the beginning charge screening and to counter ion condensation. Further addition of surfactant leads again to an increase in the surface potential, which could result from both the increased adsorption of charged molecules at the interface and the change of molecular conformation and the corresponding dipole moment. A reason for the fact that a lower potential is observed compared to the lowest investigated surfactant concentration could be due to the increasing counterion concentration which leads to a higher ionic strength in the system.[16]

Besides the non-monotonous behaviour of the surface potential, the disjoining pressure isotherms of $C_{14}TAB$ are in good agreement with the findings of other studies on cationic surfactants,[16,17] where an increasing surfactant concentration leads to thinner and more stable films. This phenomenon agrees also with the qualitative expectations on the basis of the electrostatic double-layer theory on increasing ionic strength in the solution.[97]

When polyelectrolytes are added to the system, the characteristics of the foam films change. Like in case of PAMPS/C_{14}TAB with an IEP of 10^{-4} M, the addition of low amounts of PAMPS, leads to a reduction of both the film stability and the surface potential. Furthermore, this is accompanied by a decrease in film thickness, implying that the polyelectrolytes that are now situated at the surface change the distribution of the counter ions in the Helmholtz layer as described above. Since the polyelectrolytes and surfactants are oppositely charged, both compounds form surface-active complexes and thereby release their counterions which leads to a gain in entropy. In analogy to the transition from a negatively charged air/water interface to a positively charged one after the addition of C_{14}TAB, one would expect a charge reversal due to the addition of oppositely charged polyelectrolytes which is indicated by an increase in stability and surface potential above the IEP. However, surface tension and elasticity contradict this assumption.

The surface tensions show the same characteristics as the PAMPS/C_{14}TAB system with the higher IEP, but the reduction in surface tensions is less pronounced. The lower adsorption to the surface can easily be explained by the lower amount of surface complexes in this concentration regime. As mentioned before, the surface coverage is much more dilute and, as predicted, the breaking point in the isotherm is shifted to the new IEP of 3×10^{-5} M. In a concentration regime between 10^{-5} M and 8×10^{-5} M PAMPS, no stable films can be formed at all. In this broad regime around the IEP, the surface coverage changes significantly. A maximal coverage is found below 3×10^{-5} M and a reduced amount of material at surface is measured above this concentration. This indicates that the surface coverage plays only a minor role in stabilising the foam film and that the net charge of the dissociated polyelectrolyte chain is more important in this case. In the described concentration range, the surface charge is obviously too low to stabilise a foam film and a sufficiently high charge in the system is not reached until the addition of 10^{-4} M PAMPS. At this point, stable foam films are formed again with a stability of almost 1000 Pa at 10^{-3} M PAMPS. This is more stable than the pure surfactant film, but less stable compared to PAMPS/C_{14}TAB investigated before, although the polyelectrolyte concentration is the same. However, this phenomenon is not reflected in the surface potentials since they are in the same range as those of foam films of PAMPS/C_{14}TAB with an IEP of 10^{-4} M. From this, the conclusion can be drawn that the stabilisation of foam films is affected by two individual parameters, the initial surfactant concentration and the polyelectrolyte/surfactant ratio in the system.

6.3.2 Effect of a lower degree of polymer charge

As mentioned before, the effect of PAMPS 60%/C_{14}TAB mixtures is even more complex. Additional to the broader destabilisation range, the surface tension and elasticity measurements reveal characteristics that differ significantly from the typical polyelectrolyte/surfactant curves

6.3 Discussion

discussed in chapter 4 and 5. On both sides of the IEP, namely between 10^{-5} M and 10^{-4} M PAMPS 60% and between 10^{-4} M and 10^{-3} M, respectively, a plateau in both the surface tension and the elasticity can be observed. In the surface tension measurements, the plateau below the IEP is significantly lower than that above this point, but the surface tension does not go back to the value of the pure surfactant as it does in case of a polyelectrolyte with 100 % charged monomer units. This means, although material is released from the surface, polyelectrolyte/surfactant complexes remain adsorbed at the interface. These findings are supported by surface elasticity measurements. Below 10^{-5} M PAMPS 60%, the elasticity has a medium value, which suggests that two compounds are adsorbed to the surface, namely polyelectrolyte/surfactant complexes and unbound surfactant molecules. Above a polyelectrolyte concentration of 10^{-5} M the surface elasticity is higher, which implies a reduced amount of unbound surfactant, but the constant values suggest that conformation at the surface stays the same even when more polyelectrolytes are added. Above the IEP of 1.6×10^{-4} M, a sudden change takes place at the surface and the adsorbed material is reduced. However, the measured values remain again constant over a concentration range of one order of magnitude and only the variation between the different frequencies changes. Both surface tension and elasticity do not go back to the value of the pure surfactant, which means that material remains at the surface.

Since PAMPS 60% has both negatively charged and nonionic monomer units, the interactions between the polymer and the surfactant are very complex. They can arise from electrostatic attraction between the oppositely charged compounds, but also from interactions between the surfactant and the nonionic unit. Adsorption of charged surfactants and nonionic polymers are widely known in literature[94,103] and are usually based on hydrophobic interactions or the formation of hydrogen bonds. For the occurrence of hydrophobic interactions, a surfactant with a sufficiently long alkyl chain is needed and the threshold is reported to be more than 12 C-atoms in the hydrophobic tail in the C_nTAB series.[94]

The strongly reduced surface tension indicates that the polyelectrolyte/surfactant complexes at the surface are more hydrophobic than complexes of 100 % charged polyelectrolytes. The increased surface-activity arises from the lower degree of charge, which makes the polymer less water-soluble. Although PAMPS is supposed to be a hydrophilic polyelectrolyte, it gains a more hydrophobic character if the degree of charge is reduced. For example, at a charge fraction below 20 %, the solubility of the polymer is significantly reduced.[52,104] Additionally, the surface elasticity measurements reveal a strong interaction between the surfactant and the polyelectrolyte even at low polymer concentration. However, the increased adsorption can not arise from hydrophobic interactions between surfactant and polymer in the complexes. In that case, the alkyl chain of the surfactant would interact with the polymer backbone, while the charged head group would be exposed to the surrounding water phase, which would make the complex more hydrophilic and not surface-active. O'Driscoll et al.[96] observed an interaction between the nonionic monomer unit and the charged head group of the surfactant, which is more suitable for this discussion. The described effect arises from an interaction between the dipole of the nonionic monomer unit and the surfactant, which leaves the hydrophobic tail of the surfactant exposed to the surrounding. Together with the more hydrophobic character of the polymer, these interactions could be responsible for the increased surface activity of the complexes.

Additionally, the polymer complexes form coils with a small hydrodynamic radius[105] when all charges are complexed, since no repulsion occurs between the polymer segments. In this case, a highly ordered polyelectrolyte/surfactant film can arise[96] at the interface. This is supported by a relatively high surface elasticity of 60 mN/m, which gives rise to the assumption that the complexes form an interconnected network at the surface. The interaction between the nonionic monomer units and the surfactant can lead to the shielding of the charged head groups. The shielded surfactant can not contribute to the surface charge anymore, which is therefore reduced. As described in the previous section, a certain amount of unbound surfactant is needed to stabilise a CBF. In case of PAMPS 60%/C_{14}TAB, this prerequisite seems not to be given in a broader concentration range, because of the head group shielding and the films are already destabilised at low polyelectrolyte concentration of 5×10^{-5} M. Additionally, aggregates are formed in the bulk at 10^{-4} M PAMPS 60%, indicating that close to the nominal IEP, 1:1 complexes occur that precipitate from the solution. However, this is not reflected in the surface tension or elasticity, which means that the aggregates do not adsorb at the interface and, hence, do not influence the surface properties.

The surface characterisation methods provide no information about the exact conformation of the complexes at the surface. At high surfactant concentrations, the aggregates, that are formed between the anionic/nonionic copolymer and the cationic surfactant in the bulk are described as micelles that are wrapped with polymers in a way that the charged units of the polymer[103] are complexed with the surfactant head groups. The nonionic monomer units are exposed to the surrounding water phase. Now, several conformations of the complexes at the surface are possible. One possibility is that the bulk aggregates adsorb at the surface as they are in course of a phase separation at the interface, which is postulated for example by O'Driscoll et al.[96] In that case, the micelles would remain in the core of the polymer coil with no contact to the air phase. Another possible mechanism would be that the aggregates change the conformation during the adsorption, so that the more hydrophobic surfactant tails are exposed to the air. This phenomenon is also observed for proteins that accumulate their hydrophobic domains in the core and expose them to the air when they are attached to the surface. This results in a flatly adsorbed protein at the interface.[106] The latter mechanism is more likely in this case, since no phase separation is observed for PAMPS 60%.

When the polymer concentration crosses the IEP of the system, the complexes at the interface are rearranged, as follows from the surface characterisation. Both the surface tension and the elasticity measurements indicate that a part of the polyelectrolyte complexes is released from the surface and the influence of unbound surfactant molecules increases again. Above the IEP, the concentration of the charged polymer segments exceeds that of the surfactant, so that not all charged units can be complexed by the surfactant. This affects the surface complexes in two ways. First, the internal repulsion of the likely charged polymer segments increases, resulting in a larger persistence length, $i.e.$ in larger coils. The higher hydrodynamic radius, in turn, leads to a reduced ordering at the surface[96,105] and a higher surface area per complex. Furthermore, fewer surfactant molecules decorate one polymer chain, thereby reducing the hydrophobicity of the complex. However, due to the nonionic monomer units, the complexes are still amphiphilic enough, so that they remain adsorbed on the surface. The surface tension and elasticity do not change upon polyelectrolyte addition, possibly due to a surface that is already saturated and where new polymer chains adsorb underneath, forming a thicker layer at the surface.

At polymer concentrations above 10^{-4} M, stable films are formed again. In this concentration regime, the negative charges of the polyelectrolyte strongly affect the film stability and the surfactant shielding does not play a role anymore. Comparing the film stabilities to former results, the films are not so stable, as from PAMPS/C_{14}TAB at the same concentration, which is also reflected in the surface potential. The latter is considerable lower compared to foam films from 100 % charged polyelectrolyte and can easily be explained by the fact that the amount of charged monomer units is 40 % lower at the same polyelectrolyte concentration. Hence, the repulsion that arises from the charged polymer segments is weaker and the corresponding foam films are less stable. However, compared to the foam films with a similar charge density (for exapmle 5×10^{-4} PAMPS and 10^{-3} M PAMPS 60%), the stabilities are in the same range. This supports the assumption that the stability arises from the charged polyelectrolyte in the films, no matter where they are situated, since one remarkable difference between the two systems is that in the latter case, the polyelectrolyte is adsorbed to the interface.

6.4 Conclusion

In the present chapter, the influence of the surfactant concentration and the degree of charge of a polymer on foam film properties of oppositely charged polyelectrolyte/ surfactant mixtures has been presented. For that purpose, foam film stabilities of PAMPS/C_{14}TAB with an IEP of 3×10^{-5} M and PAMPS 60%/C_{14}TAB with an IEP of 1.6×10^{-4} M has been investigated an compared to former results. In both systems, the behaviour of foam films resembles, in principal, that of the former systems, where the addition of polyelectrolytes first reduces the stability until a complete destabilisation at the IEP. Above the IEP, very stable films are formed again, even more stable than the pure surfactant foam films. However, one remarkable difference is observed between the systems: In case of a shifted IEP, the range of unstable foam films is much broader.

In a first set of experiments, the surfactant concentration which determines the IEP was reduced, which leads to a more dilute system. In the dilute system, the critical concentration, where no stable films are formed, is reached at lower polyelectrolyte concentration, although the surface tension measurements show the same characteristics as the PAMPS/C_{14}TAB described in chapter 4. This implies that below the IEP, a certain amount of unbound surfactant is needed to stabilise a CBF. Furthermore, the foam film stabilities reveal that Π_{max} strongly depends on the polyelectrolyte concentration, but also on the initial surfactant concentration.

In a second approach, the degree of charge of the polyelectrolyte was reduced to change the IEP. For this reason, a random copolymer of anionic and nonionic monomer was used, which makes the explanation of the interactions in the foam film more complex. The analysis of the results shows that the newly formed complexes are more hydrophobic. Additionally, the surfactant interacts with the dipole of the nonionic polymer segments as well. The latter leads to a shielding of the surface charge and a destabilisation of the foam film already at low polymer concentrations. Above the IEP, some of the polyelectrolyte/surfactant complexes are released from the interface, but there is still material left at the surface. Furthermore, stable films are formed again at this point, meaning that no matter where polyelectrolyte is situated in the film, it contributes to the stabilisation of the film.

In general, one can conclude that the position of the IEP does not influence the general characteristics of the film stability curve, but has a strong impact on the concentration range where films are destabilised.

7 Polyelectrolyte versus monomer effect

Abstract

So far, the influence of different oppositely charged polyelectrolytes/surfactant mixtures on the foam film stability around the IEP has been investigated. The question arises, if the effect of the polymer is related to the size of the macromolecule and to which extent it acts as a salt. For this purpose the polyelectrolyte is exchanged by the corresponding monomer. To gain a better understanding of the effect of the monomer units on the foam films, and to explore how the interactions between the different components are affected, they are compared to simple salt and surfactant systems.

The comparison between the different systems leads to the conclusion that the size of the additive indeed influences the interactions in the foam films and the film stabilities. Furthermore, the hydrophobic/hydrophilic balance of the monomer strongly changes the effect on the foam films: AMPS, the monomer of the polyelectrolyte PAMPS, behaves similar to simple salts, whereas NaSS, the monomer of the more hydrophobic PSS, shows rather the character of a cosurfactant.

7.1 Introduction

In the previous chapters, the influence of polyelectrolytes on oppositely charged surfactant foam films has been investigated. The addition of polyelectrolytes reduces the stability of the foam film towards the IEP and induces a destabilisation close to the nominal IEP of the mixture. However, the investigation of the surface properties shows that this effect in not due to a charge reversal at the interface, which has been the hitherto assumption.[90,107]

The question arises, to what extend the foam film stabilities and the surface coverage are affected by the polymeric character and if the character of the monomer units plays a role. Does the polyelectrolyte act as salt? For this purpose the polyelectrolyte is exchanged by the corresponding monomer. In that case, the ionic strength and the character of the monomer unit remain the same, but polymer character is missing. To distinguish the polymer effect from the specific effect of the monomer unit, foam films are investigated under the same conditions, including surfactant concentration, temperature, dipping time *etc.*, but with the respective monomer. Furthermore, the results are compared to measurements of foam films where other small additives were used. Since the characteristics of the monomers used in this study are different, they are compared to different model systems. To investigate the influence of the hydrophobicity on the foam film properties, two polyelectrolytes with different chemical structure were used in the former study. In case of the polyelectrolyte, the influence of hydrophobicity on foam film stabilities was rather minor. However, the different hydrophobicities of the corresponding monomers could have a major influence on the foam film stability.

In case of AMPS, the monomer of the polyelectrolyte PAMPS corresponds to a hydrophilic, monovalent ion. To distinguish the polymer effect from the salt effect, the results of AMPS/ surfactant mixtures are compared to foam films containing the monovalent salt NaCl. The addition of salt to pure surfactant solutions is already well established in the high concentration regime.[2,16,89,97,108–111] The salt ions screen the repulsion between the surfactant head groups, which leads to a shift of the cmc to lower concentrations, at least at salt concentrations above 10^{-2} M. At even higher salt concentrations, the charge screening in foam films can induce a transition from a CBF to NBF and has been observed for both cationic and anionic surfactant systems.[2,16]

In case of NaSS, the monomer of PSS has both a charged, hydrophilic sulfonate group and a more hydrophobic benzene ring. This hydrophobic/hydrophilic character makes the molecule accessible to additional hydrophobic interactions with the surfactant, so that NaSS resembles a cosurfactant. Cosurfactants can have a strong impact on the surfactant system: For example, the addition of small amounts of dodecanol to SDS solutions can induce a significant change of the morphology of the surface layer.[112,113] Furthermore, foam films from commercial C_{12}TAB samples are considerable more stable than those from the purified surfactant. The increase in film stability is due to hydrophobic impurities in the surfactant, which are usually long chained alcohols from the synthesis.[114] However, pure NaSS, like PSS, does not reduce the surface tension,[33] so that it is not sure how distinct the cosurfactant character will be.

Similar molecules to NaSS have been investigated in solutions of mixtures with C_{16}TAB. It has been shown that the physical properties of aqueous solutions are very sensitive to the molecular structure of the ion. Anions containing benzene rings induce, for example, the sphere-rod transitions of micelles at lower concentrations than in the pure surfactant system.[115,116]

In the following, results on monomer/surfactant foam film are presented and compared to the described model systems. The surfactant concentration in all experiments is 10^{-4} M which also determines the nominal IEP in all systems.

7.2 Results

C_nTAB/AMPS mixtures

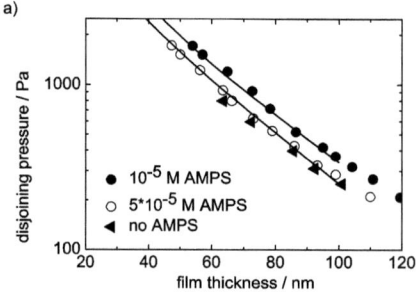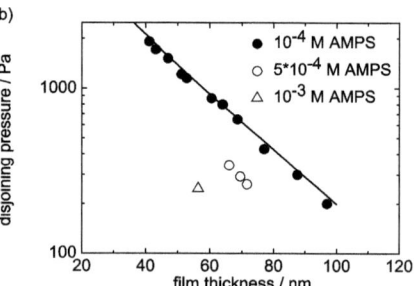

Figure 7.1: *Disjoining pressure isotherms of AMPS/C_{14}TAB solutions with varied monomer concentration; a) below the nominal IEP; b) at and above the nominal IEP; the solid lines correspond to the simulation of the isotherm with constant potential; for the sake of clarity, only some of the simulated isotherms are shown in the graph.*

In the following, the influence of AMPS, the monomer of the polyelectrolyte PAMPS, on foam film stability is investigated. Fig. 7.1 shows disjoining pressure isotherms of foam films

7.2 Results

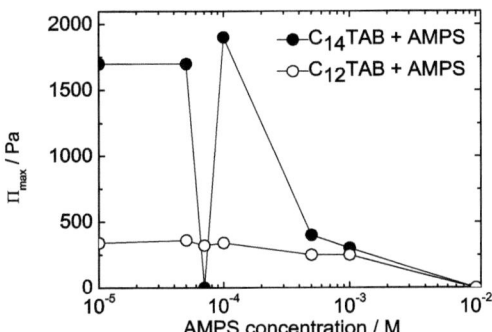

Figure 7.2: *Stability of AMPS/C_nTAB films with different surfactants; AMPS/C_{12}TAB (open circles); AMPS/C_{14}TAB (filled circles); maximum disjoining pressure Π_{max} before film rupture versus monomer concentration.*

from AMPS/C_{14}TAB mixtures at different AMPS concentrations. For the sake of clarity, the isotherms are divided into two groups. In Fig. 7.1a, disjoining pressure isotherms at concentrations below the IEP are depicted, while in Fig. 7.1b, isotherms at and above the nominal IEP are shown. At 10^{-5} M AMPS, the equilibrium thickness at a disjoining pressure of 200 Pa is around 120 nm, which is rather thick for foam films at the used ionic strength. Increasing monomer concentration leads to a strong reduction in film thickness, it decreases from 100 nm at 10^{-4} M AMPS to 71 nm at 5×10^{-4} M and 58 nm at 10^{-3} M AMPS, respectively, which is due to the electrostatic screening in the system. The equilibrium thicknesses mentioned in this paragraph refer to a disjoining pressure of 200 Pa. In addition to the film thinning, a slight increase in the slope of the isotherms is observed. This is owed to the fact that the ionic strength in the system increases from 1.1×10^{-4} M to 1.5×10^{-4} M (*cf.* Table 7.1).

For a better illustration of the foam film stability, the maximum pressure that can be applied to the film before rupture (Π_{max}) versus the monomer concentration is depicted in Fig. 7.2. The addition of 10^{-5} M AMPS leads to the formation of very stable foam films. In that case, Π_{max} is 1700 Pa, which is much higher than in case of the pure surfactant (900 Pa at 10^{-4} M C_{14}TAB). When the monomer concentration is increased to 5×10^{-5} M, no effect on the stability of the foam films can be observed. Further increase of the AMPS concentration to 7.5×10^{-5} M leads to the immediate rupture of the film after the formation, so that no stable films can be formed close to the nominal IEP of the AMPS/C_{14}TAB system. However, at the nominal IEP of 10^{-4} M, the foam films are even more stable than at low AMPS concentrations with a maximum pressure of 2000 Pa. When the monomer concentration is further increased, the foam films get less stable, so that the stability is reduced to 350 Pa at 10^{-3} M and finally to 0 Pa at 10^{-2} M AMPS. Unfortunately, only the most stable disjoining pressure isotherms give information about the surface potential of the foam films. To get the surface potential of the films, the disjoining pressure isotherms are simulated by solving the non-linear Poisson-Boltzmann equation under the assumption of constant potential. The simulations were performed with the PB program version 2.2.1[12] by Per Linse *et al.* and are summarized in Table 7.1. The addition of small amounts of monomer lead to an strong increase of surface potential to 115 mV compared to 78 mV of the pure surfactant, which explains also the increase of film thickness at this AMPS concentration. Further addition of monomer results in a slight decrease of the surface potential but it is still higher than that of the pure surfactant film. No simulations were performed at AMPS concentrations $\geq 5 \times 10^{-4}$ M, since it is not reasonable to make simulations with 3 or less data points. However, in this system, no direct correlation between film stability and surface potential can be observed.

AMPS conc. [M] (C_{14}TAB)	Ψ_0 [mV]	κ^{-1} [nm]	κ^{-1}_{theo} [nm]	I [M]
no AMPS	78	29.8	30.4	1.0×10^{-4}
1×10^{-5}	115	27.2	28.9	1.1×10^{-4}
5×10^{-5}	103	26.8	25.0	1.4×10^{-4}
1×10^{-4}	93	24.2	21.6	1.5×10^{-4}
AMPS conc. [M] (C_{12}TAB)	Ψ_0 [mV]	κ^{-1} [nm]	κ^{-1}_{theo} [nm]	I [M]
1×10^{-5}	73	30.9	28.9	8.0×10^{-5}
5×10^{-5}	75	27.3	25.0	9.0×10^{-5}
7.5×10^{-5}	70	26.8	23.1	9.0×10^{-5}
1×10^{-4}	66	24.0	21.6	1.3×10^{-4}

Table 7.1: *Summary of the surface potentials Ψ_0 from the simulation of the disjoining pressure isotherms of AMPS/C_nTAB films; the Debye length κ^{-1} is calculated from a fit of the experimental data with a exponential decay function of first order, the ionic strength I is derived from the simulation of the surface potential; κ^{-1}_{theo} corresponds the Debye length when all ion pairs are assumed to be dissociated.*

When the surfactant is changed to C_{12}TAB, the picture is slightly different. At first, it is worth to mention that the addition of AMPS leads to stable films, while foam films that are formed from pure C_{12}TAB solutions, are not stable. This is due to the shorter hydrophobic tail of the surfactant and the resulting lower adsorption coefficient[17] and is explained in more detail in chapter 5. In Fig. 7.3, two sets of disjoining pressure isotherms of AMPS/C_{12}TAB mixtures are shown. At monomer concentrations $\leq 10^{-4}$ M, the isotherms coincide on one curve in terms of thickness and differ only slightly in slope. In contrast to the AMPS/C_{14}TAB mixtures, the equilibrium thickness at a low disjoining pressure of 200 Pa is 100 nm for all 4 isotherms. However, when the monomer concentration is further increased, the film thickness is significantly reduced to 70 nm at 5×10^{-4} M and 57 nm at 10^{-3} M AMPS. As shown in Fig. 7.2, the foam film stability in the concentration regime below the nominal IEP is around 350 Pa. At 7.5×10^{-5} M AMPS, the film seems to be slightly less stable than those at 5×10^{-5} M and 10^{-4} M, respectively, but the difference is within the experimental error of the method, so that this can not be taken under consideration. With increasing AMPS concentration a decrease of film stability to 0 Pa at an AMPS concentration of 10^{-2} M is observed. Comparing these results to AMPS/C_{14}TAB, one clearly sees that below and at the nominal IEP of 10^{-4} M, the isotherms are considerable less stable than those of the latter system. However, when the monomer concentration is increased to around 5×10^{-4} M both stability curves almost coincide, as shown in Fig. 7.2. The surface potentials obtained from the simulation of the respective isotherms are rather constant at 70 ± 5 mV, which is also reflected in the constant stability of the foam films up to 10^{-4} M. As in the former mixture, no simulations were made for the isotherms with a higher monomer concentration, since there are not enough data points.

To gain a deeper insight into the surface characteristics, surface tension measurements were carried out. As shown in Fig. 7.4, the addition of small amounts of AMPS to C_{14}TAB solutions reduces the surface tension of the solution about 3 units to 67 mN/m and is constant up to an AMPS concentration of 10^{-4} M. At monomer concentrations larger than 10^{-4} M, the screening between the surfactant head groups sets in and the surface tension is strongly reduced to 51

7.2 Results

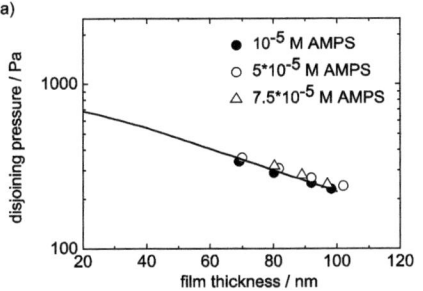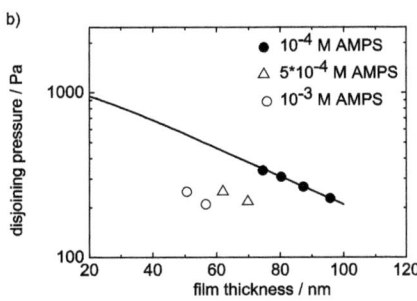

Figure 7.3: *Disjoining pressure isotherms of $AMPS/C_{12}TAB$ solutions with varied monomer concentration; a) below the nominal IEP; b) at and above the nominal IEP; the solid lines correspond to the simulation of the isotherm with constant potential; for the sake of clarity, only some of the simulated isotherms are shown in the graph.*

mN/m at 10^{-2} M. In case of $C_{12}TAB$, the decrease of surface tension in the low concentration regime is with 69 mN/m not so distinct. Furthermore, the screening starts at concentrations one order of magnitude above that of $AMPS/C_{14}TAB$, which can be explained by the lower adsorption of $C_{12}TAB$ to the surface.

$C_n TAB$/NaCl mixtures

To get a better understanding of the foam films from the monomer AMPS with C_nTAB, it is useful to compare them with foam films from mixtures of C_nTAB with a simple salt. In Fig. 7.5, disjoining pressure isotherms of $C_{14}TAB$ with NaCl are shown. Again, they are divided into two groups: isotherms below the point of equal concentrations (a) isotherms at and above 10^{-4}M NaCl (b). Up to a concentration of 10^{-4} M NaCl, the isotherms coincide on one curve with an equilibrium thickness of 100 nm at a disjoining pressure of 250 Pa. With further increase of salt concentration, the equilibrium thickness is reduced to 73 nm at 5×10^{-4} M and 62 nm at 10^{-3}M NaCl. The slope of the isotherms is very similar in the low concentration regime and first starts to change significantly at a concentration of 5×10^{-4} M.

The addition of small amounts of NaCl to $C_{14}TAB$ solutions leads to foam films with a maximum pressure of 1000 Pa, which is similar compared to the respective pure surfactant film. The stabilising effect of NaCl on foam films is not so pronounced than in case of AMPS addition but it follows the same trend (*cf.* Fig. 7.6). At 7.5×10^{-5} M NaCl, a minimum in foam film stability of 400 Pa is observed. The minimum stability is observed at the same concentration as the complete destabilisation in the $AMPS/C_{14}TAB$ system. However, at a disjoining pressure of 380 Pa, small black dots appear on the film surface and lead to the rupture of the film after a few minutes. This is a hint for the formation of quite unstable foam films and as a consequence, the film is considerably less stable at this concentration. When the NaCl concentration is further increased to 10^{-4} M, foam films with a stability of 1000 Pa can be formed again. The surface potentials in this concentration regime between 10^{-5} M and 10^{-4} M NaCl support these findings since they are rather constant within a range of 85 ± 3 mV (*cf.* Table 7.2). The surface potential of the foam film with a salt concentration of 7.5×10^{-5} M can not be simulated, since only 3 data points are available. At higher salt concentrations, the stability decreases steadily from a Π_{max} of 600 Pa for 5×10^{-4} M NaCl, and 400 Pa for 10^{-3} M to a

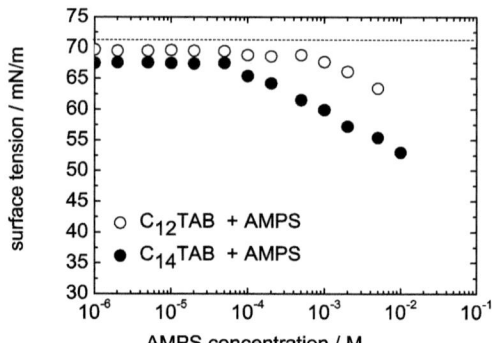

Figure 7.4: *Surface tension of AMPS/C_nTAB solutions with fixed surfactant (10^{-4} M) and variable AMPS concentration; AMPS/C_{12}TAB (open circles); AMPS/C_{14}TAB (filled circles); the dashed line corresponds to the surface tension of the pure surfactant.*

complete destabilisation at 10^{-2} M. In contrast to the destabilisation close to the IEP, the film rupture is caused by the screening of the surfactant charges and the reduced repulsion between the opposing interfaces. This is visible in a continuous thinning of the foam film upon film formation which is accompanied by the film rupture at film thicknesses around 10 nm. Due to the destabilisation of the foam films, only the film at 5×10^{-4} M NaCl could be simulated. The potential of 57 mV is considerably lower than that at 10^{-4} M.

When the surfactant is changed to C_{12}TAB, the properties of the foam films are slightly different, which is manifested in the respective disjoining pressure isotherms in Fig. 7.7. The equilibrium thickness at 250 Pa in this concentration regime, is with 105 nm in the same range as the former mixture. The only exception is the foam film at 7.5×10^{-5} M NaCl, which is slightly thicker. As described earlier, the film thicknesses are reduced and the slope of the isotherms increases in the concentration regime $\geq 5 \times 10^{-4}$M. The film stabilities shown in Fig. 7.6, below the point of equal concentrations are with a Π_{max} of 500-600 Pa slightly lower than in case of NaCl/C_{14}TAB. However, the addition of NaCl to the surfactant solution makes it possible to form stable films. As mentioned above, pure C_{12}TAB solutions do not form any stable film in this concentration regime. In contrast to all other mixtures, the increase of salt concentration to 5×10^{-5} M enhances the stability of the foam film compared to that at 10^{-5} M. Additionally, no destabilisation or minimum stability can be observed at 7.5×10^{-5} M NaCl in case of NaCl/C_{12}TAB mixtures. Further addition of NaCl leads to the formation of foam films with a stability of 1000 Pa. Furthermore, at salt concentration $\geq 10^{-4}$ M, the disjoining pressure isotherms of both NaCl/surfactant mixtures almost coincide in terms of stability and film thickness. This indicates that the ionic strength and the resulting charge screening plays an important role. The surface potentials, calculated from the respective isotherms, of 80 ± 5 mV below and at the nominal IEP are very similar to those of the NaCl/C_{14}TAB mixtures. The comparison of the surface potentials at higher concentrations is difficult, since due to the above described problems, only the isotherm at 5×10^{-4} M NaCl could be simulated. However, the potential of 48 mV at this concentration is only slightly lower than that of the film from the respective NaCl/C_{14}TAB solution.

The surface tension isotherms of NaCl/C_nTAB show similar trends as AMPS/C_nTAB mixtures

7.2 Results

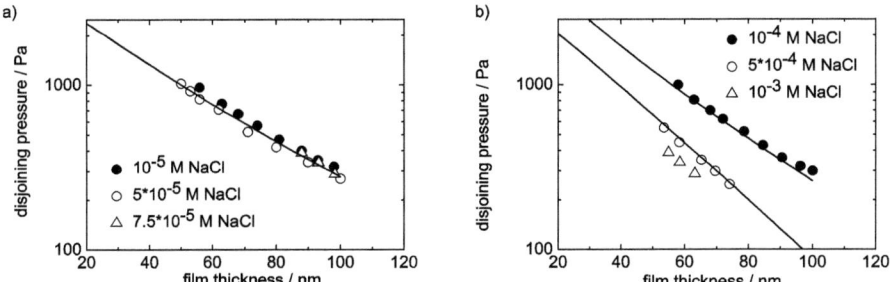

Figure 7.5: *Disjoining pressure isotherms of $NaCl/C_{14}TAB$ solutions with varied salt concentration; a) below the point of equal concentrations; b) at and above the point of equal concentrations. The solid lines correspond to the simulation of the isotherm with constant potential; for the sake of clarity, only some of the simulated isotherms are shown in the graph.*

(*cf.* Fig. 7.8). However, the screening starts at higher concentrations than in case of AMPS, namely at 10^{-3} M for C_{14}TAB and 10^{-2} for C_{12}TAB, respectively, which is one order of magnitude higher than in the former case. Additionally, the reduction of surface tension for C_{14}TAB in the low salt concentration regime is not so pronounced so that the surface tension is only slightly lower than for mixtures with C_{12}TAB. The isotherms of both mixtures are very close to the surface tension of the pure surfactant solution in this concentration range.

C_nTAB/NaSS mixtures

In the following, results of NaSS/C_nTAB mixtures are presented. In Fig. 7.9, disjoining pressure isotherms of NaSS/C_{14}TAB mixtures are shown. The addition of small amounts of the monomer NaSS (10^{-6} M) to a 10^{-4} M C_{14}TAB solution leads already to the formation of very stable films compared to the pure surfactant film (Π_{max} = 1250 Pa vs. 900 Pa for the pure surfactant). When the NaSS concentration is increased to 10^{-5} M, the stability is reduced. Both isotherms coincide, indicating that the ionic strength at such low monomer concentrations is dominated by the surfactant. Indeed, the foam films are very similar to the pure C_{14}TAB film (*cf.* chapter 4) in terms of ionic strength, surface potential and equilibrium thickness. When the monomer concentration is further increased, no stable foam films can be formed in a broad concentration regime (*cf.* Fig. 7.11). At NaSS concentrations as high as 10^{-3} M, a stable film is observed at low disjoining pressures. A few minutes after formation, a black dot appears that spreads very slowly (see Fig. 7.16 left). The film is stable for hours, without completing the transition to a NBF. Only when the applied pressure is increased, further spreading is induced and the film ruptures eventually. When the monomer concentration is further increased, aggregates are formed between the two components due to the hydrophobic interaction between NaSS and C_{14}TAB. An inhomogeneous film can be formed and a CBF-NBF transition can be observed immediately (see Fig. 7.16 right). However, after a few minutes, the film ruptures due to the aggregates that are trapped in the film.

Foam films of NaSS/C_{12}TAB mixtures are a rather similar compared to the systems with C_{14}TAB and are shown in Fig. 7.10. The film thicknesses at a disjoining pressure of 200 Pa at concentrations below the IEP are in the same range than those of NaSS/C_{14}TAB. At very

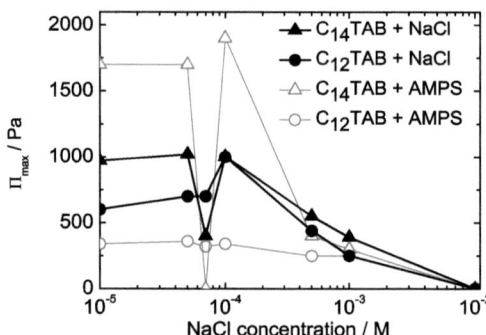

Figure 7.6: *Stability of NaCl/C_nTAB films with different surfactants; NaCl/C_{12}TAB (filled circles); NaCl/C_{14}TAB (filled triangles); maximum disjoining pressure Π_{max} before film rupture versus salt concentration; for comparison, the stabilities of AMPS/C_nTAB systems are shown in gray.*

low concentrations, stable films can be formed in contrast to pure surfactant films that are not stable at all. Compared to NaSS/C_{14}TAB, the absolute stabilities of 1000 Pa at 10^{-6} M and 300 Pa at 10^{-5} M NaSS, respectively, are reduced (see Fig. 7.11), which can be explained by the lower surface coverage of surfactant molecules in case of C_{12}TAB films.[17] This is also reflected in the surface potential of 76 mV at 10^{-6} M, which is slightly lower than that of the corresponding foam film with C_{14}TAB. In the concentration regime between 10^{-5} M and 10^{-2} M NaSS, no stable films are formed at all, but when the monomer concentration is increased to 10^{-2} M, stable foam films are observed again. Above a disjoining pressure of 200 Pa a CBF-NBF transition takes place and the black domains spread over the whole film within seconds. This transition is shown in Fig. 7.17 (left). When the transition is completed, the film ruptures immediately. In this concentration regime, the charge screening plays an important role, so that the films are considerably thinner compared to the foam films at lower concentrations. At a monomer concentration of 10^{-1} M, a transition to a NBF occurs directly after film formation and the resulting NBF with a thickness of 7 nm is stable beyond the maximum pressure that is applicable in the experiment. The occurrence of NBF is very similar to the NaSS/C_{14}TAB system, but takes place at concentrations one order of magnitude higher. In this concentration regime, no simulations of the surface potential can be performed. However, this type of film is stabilised by steric forces and not by electrostatic forces, so that the surface potential is only secondary.

The effect of the NaSS on the surface tension isotherms is rather complex. In a first set of experiments, the influence of a fixed amount of NaSS on the surface tension of the surfactant solutions is investigated. These changes in the isotherms upon addition of 10^{-3} M NaSS to C_nTAB solutions are shown in Fig. 7.12a. For better comparison, the surface tension isotherms of the pure surfactant are depicted as well. The addition of NaSS to the surfactant solutions results in an strong decrease of the surface tension, and this effect is more pronounced for NaSS/C_{14}TAB. Below the respective cmc of the surfactant a minimum of 30 mN/m is reached. This value is identical in both cases; only the surfactant concentration at which this minimum occurs, is different. Further addition of the surfactant leads to an increase of surface tension and both curves coincide on that of the pure surfactant, which is in good agreement with a

7.2 Results

NaCl conc. [M] (C_{14}TAB)	Ψ_0 [mV]	κ^{-1} [nm]	κ^{-1}_{theo} [nm]	I [M]
1×10^{-5}	87	26.1	28.9	9.0×10^{-5}
5×10^{-5}	82	25.5	25.0	1.0×10^{-4}
1×10^{-4}	85	23.6	21.6	1.4×10^{-4}
5×10^{-4}	57	17.2	12.4	2.1×10^{-4}
NaCl conc. [M] (C_{12}TAB)	Ψ_0 [mV]	κ^{-1} [nm]	κ^{-1}_{theo} [nm]	I [M]
1×10^{-5}	85	30.5	28.9	8.0×10^{-5}
4×10^{-5}	77	28.5	25.9	9.0×10^{-5}
7.5×10^{-5}	80	27.3	23.1	1.1×10^{-4}
1×10^{-4}	80	23.7	21.6	1.4×10^{-4}
5×10^{-4}	48	17.9	12.4	2.0×10^{-4}

Table 7.2: *Summary of the surface potentials Ψ_0 from the simulation of the disjoining pressure isotherms of NaCl/C_nTAB films; the Debye length κ^{-1} is calculated from a fit of the experimental data with a exponential decay function of first order, the ionic strength I is derived from the simulation of the surface potential; κ^{-1}_{theo} corresponds the Debye length when all ion pairs are assumed to be dissociated.*

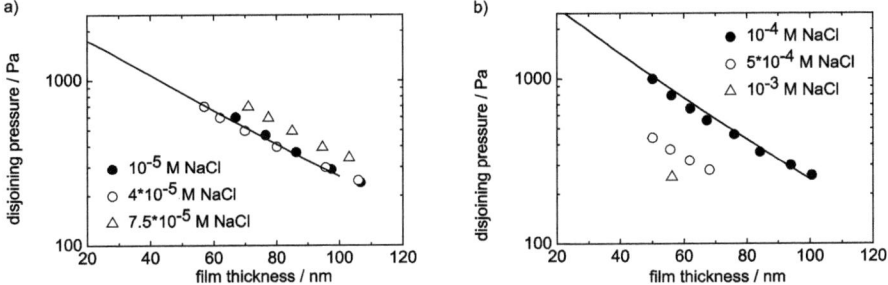

Figure 7.7: *Disjoining pressure isotherms of NaCl/C_{12}TAB solutions with varied salt concentration; a) below the point of equal concentrations; b) at and above the point of equal concentrations; the solid lines correspond to the simulation of the isotherm with constant potential; for the sake of clarity, only some of the simulated isotherms are shown in the graph.*

NaSS conc. [M]	Ψ_0 [mV]	κ^{-1} [nm]	κ^{-1}_{theo} [nm]	I [M]
1×10^{-6} + C_{14}TAB	83	27.9	30.2	1.0×10^{-4}
1×10^{-6} + C_{12}TAB	76	26.9	30.2	1.0×10^{-4}

Table 7.3: *Summary of the surface potentials Ψ_0 from the simulation of the disjoining pressure isotherms of NaSS/C_nTAB films; the Debye length κ^{-1} is calculated from a fit of the experimental data with a exponential decay function of first order, the ionic strength I is derived from the simulation of the surface potential; κ^{-1}_{theo} corresponds the Debye length when all ion pairs are assumed to be dissociated.*

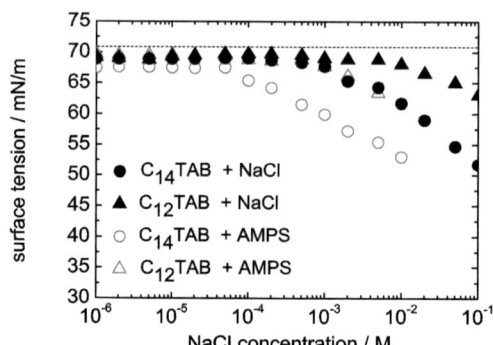

Figure 7.8: *Surface tension of NaCl/C_nTAB solutions with fixed surfactant (10^{-4} M) and variable NaCl concentration; NaCl/C_{12}TAB (filled triangles); NaCl/C_{14}TAB (filled circles); for comparison, the surface tension of AMPS/C_nTAB systems are shown in gray; the dashed line corresponds to the surface tension of the pure surfactant.*

study of Monteux et al.[33]

The surface tension at a fixed surfactant concentration in dependence of the monomer concentration is shown in Fig. 7.12b. The surface tension isotherm resembles that of the pure surfactant and no minimum is observed. The addition of NaSS to the surfactant solutions leads to a continuous reduction in surface tension until a plateau is reached. This plateau gives evidence of the existence of a cmc in the system which, in turn, indicates the formation of micelles in the solution. In contrast to the above described surface tension isotherm with varied surfactant concentration, no minimum is observed. In the case of NaSS/C_{14}TAB, the decrease in surface tension is more pronounced and the cmc is reached at lower monomer concentration of 2×10^{-3} M. Furthermore, the surface tension value of the plateau region is 30 mN/m and therefore considerably lower than for NaSS/C_{12}TAB mixtures. In this case, the cmc is observed at 2×10^{-2} M at a surfaces tension of 35 mN/m. Both cmc values are in the same concentration range as those of the pure surfactant but the plateau values are lower. Considering the cmc, it is worth mentioning that the foam films in the concentration regime above this point show no stratification, although micelles should be present in the bulk solution.

C_{12}TAB/C_{10}SO$_3$ mixtures

Since NaSS is more hydrophobic than AMPS, it could also have the characteristics of a co-surfactant. To gain a better understanding of the monomer, measurements with the anionic surfactant C_{10}SO$_3$ were performed. These measurements are shown in Fig. 7.13. At the lowest investigated C_{10}SO$_3$ concentration, the foam film is very similar to that of NaSS/C_{12}TAB in terms of slope surface potential and equilibrium thickness (*cf.* Table 7.4). With increasing monomer concentration, both film thickness and surface potential are reduced due to the higher ionic strength and the associated screening in the system. As described earlier, C_{12}TAB forms no stable foam films, but the addition of 10^{-6} M C_{10}SO$_3$ leads to the formation of foam films with a Π_{max} of 600 Pa (see Fig. 7.14). The increase of the surfactant concentration leads to a broad regime of unstable films up to a concentration of 5×10^{-4} M. At this concentra-

7.2 Results

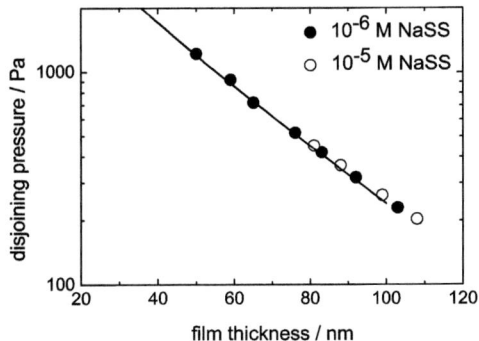

Figure 7.9: *Disjoining pressure isotherms of NaSS/C_{14}TAB solutions with varied monomer concentration, below the IEP; the solid line corresponds to the simulation of the isotherm with constant potential.*

$C_{10}SO_3$ conc. [M]	Ψ_0 [mV]	κ^{-1} [nm]	κ^{-1}_{theo} [nm]	I [M]
1×10^{-6}	80	29.4	30.3	1.0×10^{-5}
5×10^{-4}	33	14.0	12.4	2.0×10^{-4}
1×10^{-3}	31	9.0	9.2	1.0×10^{-3}

Table 7.4: *Summary of the surface potentials Ψ_0 from the simulation of the disjoining pressure isotherms of $C_{10}SO_3/C_{12}TAB$ films; the Debye length κ^{-1} is calculated from a fit of the experimental data with a exponential decay function of first order, the ionic strength I is derived from the simulation of the surface potential; κ^{-1}_{theo} corresponds the Debye length when all ion pairs are assumed to be dissociated.*

tion, a foam film with a stability of 350 Pa can be formed. The effect is comparable to that of NaSS/C_{12}TAB but the stability increase at low concentrations is less pronounced. Additionally, the critical concentration at which stable foam films can be formed again, is considerably lower than in case of NaSS mixtures. When the surfactant concentration is increased, the stability of the foam films strongly increases to 2800 Pa at 10^{-3} M and 6800 Pa at 10^{-2} M $C_{10}SO_3$. At 0.1 M $C_{10}SO_3$, stratification due to micelles is observed at low disjoining pressures and finally, a CBF-NBF transition takes place (*cf.* Fig. 7.18).

This NBF has a thickness of 10 nm and was only stable up to a disjoining pressure of 200 Pa (see Fig. 7.14). In contrast to that, the NBF of the NaSS/C_{12}TAB system at this concentration was stable and no stratification could be observed before the NBF transition.

The shape of the surface tension isotherm is similar to that of NaSS/C_{12}TAB mixtures of a fixed monomer concentration (see Fig. 7.12a). However, the addition of $C_{10}SO_3$ leads to a stronger decrease at low concentrations ($< 10^{-4}$ M) and the cmc appears at one order of magnitude lower. The minimum of 30 mN/m has the same value, only the surface tension of the plateau is slightly higher (37 mN/m), which is very similar to that of the $C_{10}SO_3$ isotherm. The stratification of the foam films start at a concentration far above the cmc at 10^{-1} M.

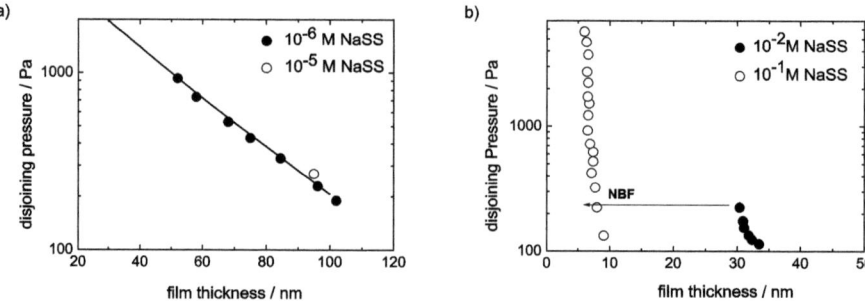

Figure 7.10: *Disjoining pressure isotherms of NaSS/C_{12}TAB solutions with varied monomer concentration; a) below the IEP; b) above the IEP; the solid line corresponds to the simulation of the isotherm with constant potential; for the sake of clarity, only some of the simulated isotherms are shown in the graph.*

7.3 Discussion

As shown in chapter 4 and 5, the addition of polyelectrolyte has a strong influence on foam film behaviour around the IEP. To distinguish between the effect of the monomer segment properties (salt effect and molecule specific effects) and the polymeric effect, PAMPS is exchanged by its monomer AMPS and foam films are investigated at the same monomer concentration.

7.3.1 Salt effect

Since AMPS is a rather small molecule, it is compared to systems with the simple salt NaCl. Indeed, the mixtures show significant similarities. In all four systems, stable foam films are formed at low monomer and salt concentrations, respectively. In case of C_{14}TAB, a destabilisation or a minimum stability can be observed in both mixtures close to the IEP, although the effect is more distinct for AMPS. Above this minimum, stable films are formed again, but the stability decreases up to a concentration of 10^{-2} M, where no stable films can be formed at all. For C_{12}TAB the effect is similar, but no minimum in stability is detected. Above a certain concentration, the film thicknesses are the same and also the destabilisation in the high concentration regime occurs at the same point for all systems. However, at concentrations $\leq 10^{-4}$ M, the films that are formed with C_{12}TAB are considerably less stable than those of C_{14}TAB mixtures.

Since AMPS is bulkier and slightly more hydrophobic than NaCl, due to the C-C double bond in the molecule, it is assumed that AMPS has a stronger tendency to penetrate into the interface of the foam film.[117] In case of C_{14}TAB, this increases the lateral screening of the charged surfactant head groups and allows the adsorption of more surfactant molecules at the surface. It is also visible in the surface tension measurements, where the surface coverage at low AMPS concentration is rather high compared to the other mixtures and the lowering of the surface tension due to charge screening is observed one order of magnitude lower. However, the lateral screening does not reduce the surface potential of the foam films. The opposite is the case, the potential increases due to the higher surfactant adsorption which results in rather thick foam films. At higher monomer concentrations, ion pairing between the surfactant and the monomer reduces the surface potential and the charge screening in the electrostatic double-layer leads to

7.3 Discussion

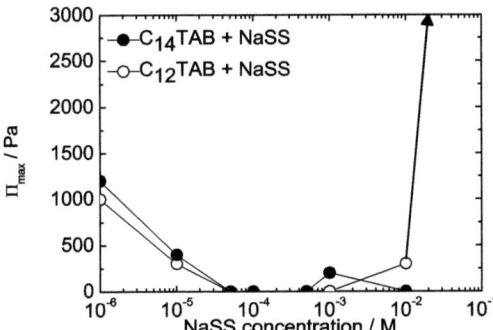

Figure 7.11: *Stability of NaSS/C_nTAB films with different surfactants; NaSS/C_{12}TAB (open circles); NaSS/C_{14}TAB (filled circles); maximum disjoining pressure Π_{max} before film rupture versus monomer concentration.*

an reduction in film thickness.

For both systems with C_{14}TAB, a destabilisation and a stability minimum, respectively is observed close to the nominal IEP and the question arises, why this happens. In case of AMPS, the formation of ion pairs between AMPS and the surfactant could be a possible reason, since the release of the smaller counterions is entropically more favourable.[27] The ion pair formation would lead to a reduced charge in the system and therefore to destabilisation of the foam film. However, this explanation fails, when systems at lower AMPS concentrations are considered. If ion pair formation occurred, the surface potential would be much lower and the foam film stability would be reduced approaching the IEP. Since this is not the case, the effect that causes the destabilisation has to be of different origin. Evidence comes from NaCl/C_{14}TAB foam films at 7.5×10^{-5} M, where small black dots appear at the film surface before film rupture. These black dots are a hint for a beginning CBF-NBF transition, implying that local heterogeneities develop due to the lack of repulsion between the interfaces and a transition starts. Stable NBF occur only, when the surface coverage is maximum and crystalline ordering of the molecules is observed.[2] Since this requirement is not fulfilled, the foam films rupture immediately when the dots spread. The same mechanism is assumed to take place in AMPS/C_{14}TAB mixtures as well, but as the screening seems to be more pronounced for AMPS, the formation of local discontinuities occurs much faster and at much lower disjoining pressures, so that the foam films rupture directly after film formation.

The described lateral screening in the low concentration regime is not so pronounced in case of C_{12}TAB. The surface tension indicates that the surface coverage is lower for AMPS/C_{12}TAB than for mixtures with C_{14}TAB. This means that even when the repulsion of the surfactant head groups is reduced, the surface layer is assumed to be very dilute. In this case, the charge screening that affects the surface tension, starts at about one order of magnitude higher than for C_{14}TAB. The effect of AMPS/C_{12}TAB and NaCl/C_{12}TAB on the surface coverage is rather similar. The surface potentials of the two systems are slightly lower than those for C_{14}TAB, which supports the hypothesis of a lower surface coverage compared to the latter systems. The dilute surface layer also explains, why the destabilisation effect that takes place in both systems with C_{14}TAB can not be observed for C_{12}TAB, since the foam films are not so sensitive to local heterogeneities.

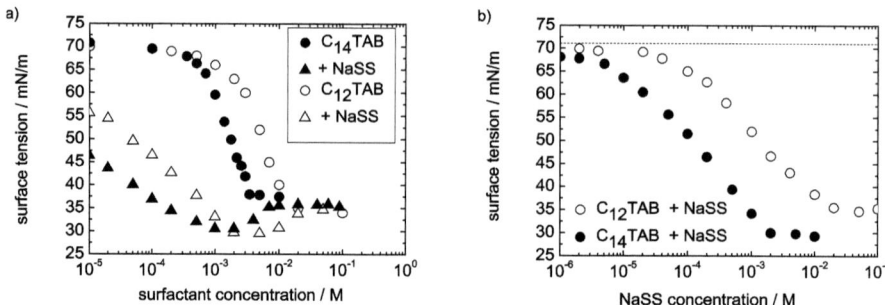

Figure 7.12: *Surface tension of NaSS/C_nTAB solutions; a) with fixed monomer (10^{-3} M) and variable surfactant concentration; b) with fixed surfactant (10^{-4} M) and variable NaSS concentration, the dashed line corresponds to the surface tension of the pure surfactant; C_{12}TAB systems (open symbols); C_{14}TAB systems (filled symbols).*

Above the point of equal concentration, the foam film properties of the different systems are very similar although the reduction in surface tension is different in all cases. In all 4 systems, the reduction of film thickness and film stability takes place in the same concentration regime. Pronounced changes in Π_{max} and the thickness are first observed at a concentration of $\geq 5\times10^{-4}$ M and the complete destabilisation occurs at 10^{-2} M. Furthermore, at the concentration of 10^{-3} M, the maximum pressure of all systems almost coincides. However, at least for the system with NaCl and AMPS/C_{12}TAB this is in a concentration range where according to the surface tension measurements the surface coverage is still very low. This means that the coverage of the film interfaces has no influence on the film thickness and stability. Therefore, one can conclude that the ionic strength, which is the same in all systems, is the crucial parameter concerning film thickness and stability.

In the high concentration regime, the addition of salt has usually two effects: First, it increases the adsorption density of ionic surfactant at the interface due to an enhanced lateral screening of the surfactant head groups and the resulting lower repulsion between the molecules at the surface.[111] This leads also to a lowering of the cmc in the system. Additionally, because of the charge screening, the salt reduces the repulsion between the two film interfaces, which leads to thinner films. Usually, when the salt concentration is sufficiently high, the surfactant head groups are screened so well that the electrostatic barrier between the film interfaces is overcome and van der Waals attraction sets in, which results in further thinning of the film. At very low distances between the two surfaces, a steric repulsion can stabilise the film again and a NBF occurs. The described phenomenon has been observed, for example, for films with C_{14}TAB or with SDS and NaCl in the concentration regime around 1 M.[16,89] However, for the formation of a stable NBF, a densely covered surface is required. This is usually the case, when the concentration is close to the cmc of the system. In case of C_nTAB at 10^{-4} M, the surface coverage is still very low. In general, at low surfactant concentrations, the amount of salt[2,118] that is needed to induce a NBF transition is even higher than in case of a nearly fully covered surface. Hence, the salt concentration that is needed to induce a transition to a NBF is very high in this system so that no NBF can be observed.

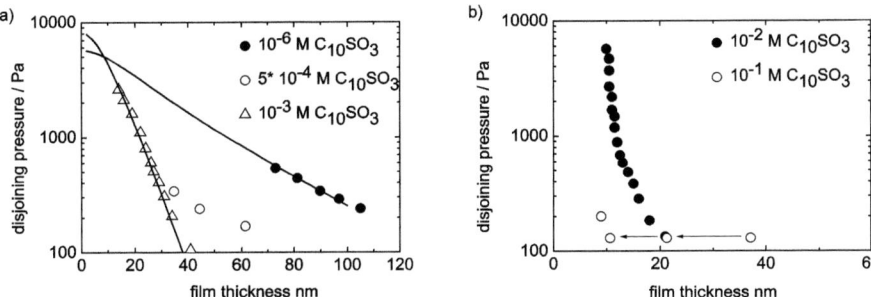

Figure 7.13: *Disjoining pressure isotherms of $C_{10}SO_3/C_{12}TAB$ solutions with varied $C_{10}SO_3$ concentration; a) at concentrations below and at 10^{-3} M; b) at concentrations above 10^{-3} M; the solid lines correspond to the simulation of the isotherm with constant potential; for the sake of clarity, only some of the simulated isotherms are shown in the graph.*

7.3.2 Hydrophilic/hydrophobic balance

In case of NaSS, the effect on foam films is quite different. The addition of very low concentrations of NaSS leads to a significant increase in film stability. This occurs already at 10^{-6} M, while in this concentration regime, the influence of the polyelectrolyte is still negligible. When the monomer concentration is further increased, the film stability is rapidly reduced. The effect is very pronounced, so that at concentrations above 10^{-5} M, no stable films are observed. The addition of the monomer leads to a very broad concentration regime, where no stable films can be formed. For $C_{12}TAB$, this regime is even larger than in case of $C_{14}TAB$. Above a concentration of 10^{-3} M for $C_{14}TAB$ and 10^{-2} M for $C_{12}TAB$, respectively, foam films get stable again. Additionally, CBF-NBF transitions are observed in this concentration regime for both systems.

The comparison between NaSS systems and those of AMPS shows significant differences as well. In fact, the monomer NaSS has more or less the opposite effect on film stabilities. It stabilises foam films at very low concentrations, where AMPS has no influence on the foam films and leads to a destabilisation in a range, where the other monomer forms very stable films. Furthermore, in the concentration regime above the IEP, very stable NBFs are formed, but in case of the AMPS, a reduction in film stability occurs due to charge screening. The size and the charge of the two monomers are comparable, which implies that properties of NaSS are dominated by the hydrophobic part of the molecules. Only in the high concentration regime, the transitions to a NBF give a hint that NaSS can also act like a salt. In general, the film stabilities of the $NaSS/C_{14}TAB$ and the surface tension isotherms give evidence, that the monomer acts like a cosurfactant. For this reason, the properties of the films with the monomer are compared to a catanionic system, consisting of the anionic surfactant $C_{10}SO_3$ and the cationic surfactant $C_{12}TAB$.

For the discussion of the foam film stabilities of mixtures with NaSS, it is important to consider the surface characteristics. The surface tension isotherms of $NaSS/C_nTAB$ mixture are quite complex. In case of a fixed NaSS concentration and a variable surfactant concentration, a strong reduction in surface tension at low surfactant concentration is observed as well as a huge minimum below the cmc, which is in good agreement with findings from Monteux et al..[33] The strong reduction in surface tension gives evidence, that the monomer is incorporated into

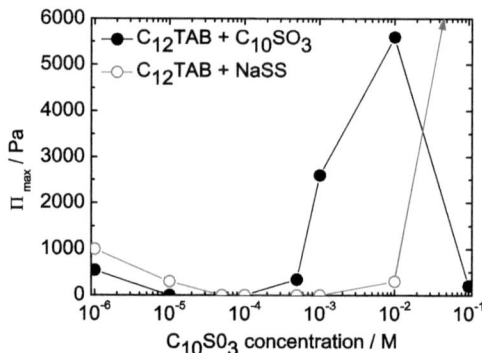

Figure 7.14: *Stability of $C_{10}SO_3/C_{12}TAB$ films; maximum disjoining pressure Π_{max} before film rupture versus $C_{10}SO_3$ concentration; for comparison the stabilities of $NaSS/C_{12}TAB$ films are shown in gray.*

the surface layer, even though pure NaSS is not surface-active. This implies that a cooperative binding process occurs when surfactant and monomer are present in the solution.[115] Furthermore, the dip in surface tension suggests that the monomer is resolubelised from the surface and integrated into micelles, so that a more or less pure surfactant layer is left at the surface. This phenomenon is also observed, when surface-active impurities are present in a surfactant solution.[119]

At a fixed C_nTAB concentration and a varied NaSS concentration, the surface tension shows a plateau at high monomer concentrations, which implies that micellar structures are formed in the bulk solution. This has also been observed for C_{16}TAB mixtures with tosylate,[115,116] a hydrophobic ion, which is very similar to NaSS, so that the two compounds can be easily compared. In a study by Li et al., the addition of the hydrophobic ion to the surfactant solution led to a transition from spherical to rod like micelles, even at rather low concentrations. Furthermore, at a high concentration of micelles, entangled structures were observed. The formation of micelles is quite surprising, especially at these low concentrations because the hydrophobic tail of the monomer is rather short. However, besides the usual hydrophobic interactions, Π-stacking can occur as well due to the aromatic ring in the monomer[120] with could enhance the formation of micelles.

The absence of a dip below the cmc in the NaSS/C_nTAB mixtures indicates in Fig. 7.12 that the surfactant is not desorbed from the surface, so that it is assumed that the surfactant is present in both the micelles and the surface layer. The surfactant type that is used in the mixture has a strong influence on the micelle formation. In case of C_{14}TAB, the cmc occurs at a concentration that is one order of magnitude lower than for C_{12}TAB. This phenomenon happens, even though the surfactant concentration is a 1/20 and a 1/200 of the monomer concentration, respectively, implying that already the addition of small amounts of molecules can have a strong influence on the system. Furthermore, the surface tension at the cmc is much lower in the case of C_{14}TAB, which indicates that the surface is more densely packed. In total, the difference in surface tension between the two systems is 5 mN/m, which can be explained by the lower adsorption coefficient of C_{12}TAB[17] and a resulting lower cooperative adsorption of NaSS to the surface.

7.3 Discussion

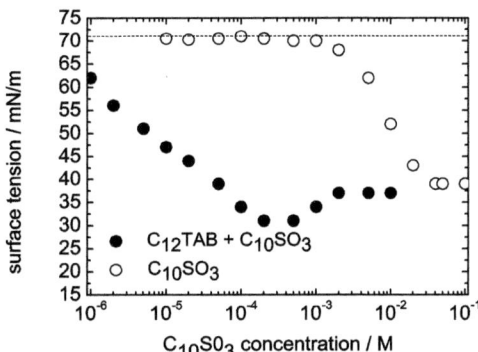

Figure 7.15: *Surface tension of pure $C_{10}SO_3$ solutions (empty circles) and $C_{10}SO_3/C_{12}TAB$ mixtures with fixed surfactant (10^{-4} M) and variable $C_{10}SO_3$ concentration (filled circles); the dashed line corresponds to the surface tension of the pure $C_{12}TAB$.*

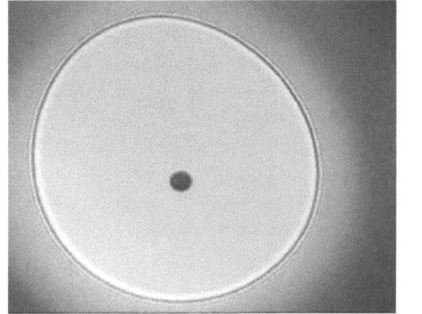 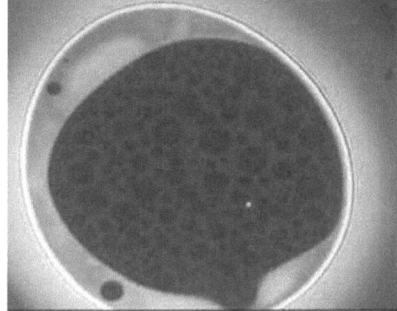

Figure 7.16: *Foam films from $NaSS/C_{14}TAB$ mixtures at a disjoining pressure of 200 Pa; left: at a NaSS concentration of 10^{-3} M; right: at a NaSS concentration of 10^{-2} M.*

In contrast to NaSS, $C_{10}SO_3$ is surface-active and can be used as a surfactant to stabilise foam films (data not shown). This is manifested in the surface tension isotherms by a much stronger reduction in surface tension and a lower cmc. An explanation for this effect is a higher adsorption coefficient of $C_{10}SO_3$ in comparison to NaSS due to the longer hydrophobic tail of the molecule. At concentrations close to the cmc, a minimum in surface tension is observed, which gives evidence that the cationic surfactant is solubilised in micelles and a pure $C_{10}SO_3$ layer is developed at the surface. In fact, the plateau value of $C_{10}SO_3$ is slightly higher, which could be because of the cooperative binding of NaSS to $C_{12}TAB$ in this concentration regime and the resulting denser surface layer.

The comparison of the film stabilities of $NaSS/C_nTAB$ mixtures with those of $C_{10}SO_3/C_{12}TAB$ shows that in the concentration range between 10^{-6} M and 10^{-5} M, foam films with NaSS are more stable. A reason for that could be the difference in cosurfactant chain length. A neutron reflectometry study of tosylate/$C_{16}TAB$ mixtures at the air-water interface showed, that small hydrophobic ions are embedded in the hydrophobic region of the surface layer.[115,121] The additional adsorption of molecules leads to the expansion of the monolayer and not to the screening of the lateral repulsion between the surfactant head groups due to ion pair formation. This, in turn, results in an increase in surface coverage rather than in a reduction of the surface charge. Foam films from mixtures with C_{14}TAB are more stable than those with C_{12}TAB, which can be explained by higher surface coverage in the first case. The simulation of the surface potentials supports these findings, since the potential of NaSS/C_{14}TAB is slightly higher than that of NaSS/C_{12}TAB at 10^{-6} M. In contrast to that, C_{12}TAB and C_{10}SO$_3$ interact stronger, since these molecules can form 1:1 complexes. According to a study on oppositely charged surfactant mixtures with different chain lengths, systems of surfactants with a similar length of the hydrophobic tail interact stronger than those with a distinct difference in chain length.[122] In the present case, due to the similar chain lengths, and the resulting chain compatibility between the two compounds, hydrophobic interaction can take place additionally to the electrostatic interactions between the head groups. This leads to a strong screening of the surface charge due to the complex formation, which has an effect on the foam films: At low C_{10}SO$_3$ concentrations, the reduced surface charge leads to foam films that become unstable close to the IEP.

In all systems, no stable films can be formed in a quite broad concentration regime. In case of C_{10}SO$_3$, the destabilisation occurs only below the IEP of the system and can be explained by the strong electrostatic interaction between the two oppositely charged compounds. This reduces the surface charge such that the repulsion between the opposing film surfaces is too low to stablise the foam films. However, when the nominal IEP is crossed, the excess of anionic surfactant leads to the formation of stable films again. In contrast to this, the addition of NaSS to the C_nTAB solutions leads to the occurrence of a destabilisation regime that is much larger, so that the destabilisation can not be related to a low surface charge. The reason for the destabilisation mechanism is rather the character of the hydrophobic ion. On one hand, hydrophobic ions are often used as defoamers, since the addition increases the distance between the surfactant molecules.[117] This can lead to a reduction of the surface potential, so that no stable films can be formed in this concentration regime. However, the destabilisation in this system takes place in a concentration regime, were the surface coverage is not very dense, so that it is not sure, if this effect can appear.

On the other hand, the hydrophobic tail of the monomer is very short compared to the used surfactant. A few studies on the influence of the surfactant chain length have shown that the surface elasticity decreases when the length of the hydrophobic tail is reduced.[17,114] The reduction in surface elasticity is usually related to a decrease in film stability and, finally, no stable films can then be obtained. This leads to the conclusion, that surface elasticity could play a crucial role in destabilising the foam films. In general, the CBFs that are formed from

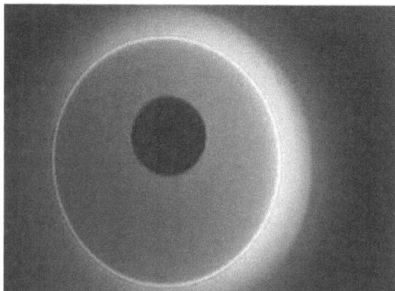 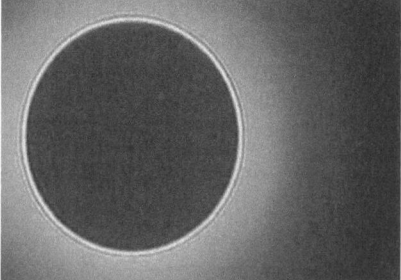

Figure 7.17: *Foam films from NaSS/C_{12}TAB mixtures; left: at a NaSS concentration of 10^{-2} M; rigth: at a NaSS concentration of 10^{-1} M.*

7.3 Discussion

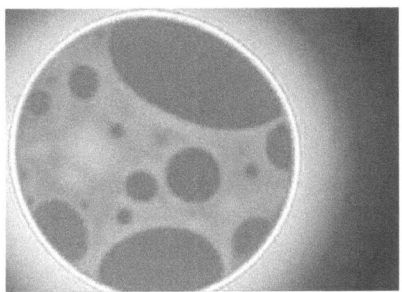 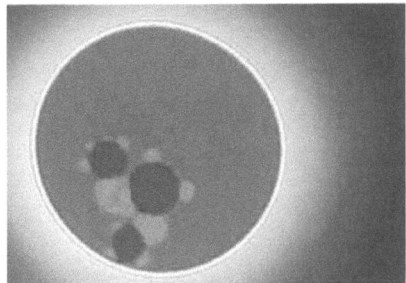

Figure 7.18: *Foam films of $C_{12}TAB/C_{10}SO_3$ mixtures at 10^{-1} M $C_{10}SO_3$; Stratification of the foam film.*

NaSS/surfactant solutions in the high concentration regime are not very stable, even though the surface tension measurements imply a high surface coverage. This supports the assumption that the surface elasticity is not high enough to stabilise the films. Only when a NBF is formed above the cmc of the system, a very stable foam film can be observed. However, the stabilisation mechanism is different for NBFs.

NaSS films do not show stratification, even though a concentration range is investigated that is about one order of magnitude higher than the cmc. This effect has been observed before by Bergeron et al.[109] and Espert et al.[111] A possible reason for this phenomenon could be that the concentration of micelles is too low close to the cmc, so that there might not be enough micelles to form ordered structures. On the other hand, the addition of salt to a solution suppresses the stratification process and enhances a transition to a NBF,[2,123,124] which can be observed in both NaSS systems. The salt character of the monomer could therefore hinder the stepwise thinning of the films above the cmc of the mixture, so that only a NBF-transition takes place. In general, micelles can act as a source for molecules that adsorb at the surface. It can be assumed, that monomers are transported to the film interface, depleted from the micelles and then integrated into the surface layer[125] so that a densely covered surface is developed, promoting the stability of the NBF.

The first CBF-NBF transition can be observed at concentrations just below the cmc of the two systems. In this case, the NBF is not very stable and ruptures shortly after the transition. As mentioned earlier, a stable NBF can only be formed, when the film surface is densely covered and the surface layer has a crystalline structure. This indicates that the surface layer has not reached the maximum density below the cmc, so that the NBF ruptures after the formation.

Usually, before a transition to a NBF occurs, small black dots appear at the films interface. The dots are often called 'holes' but are rather regions of smaller film thickness.[2] These thin domains spread over the whole film, due to a difference in interfacial tension between the two regimes. Since the surface tension is lower in the inner part of the domains and a lower energy state in the system is more favorable, the spreading is enhanced and a new, thinner film is formed. In case of a NaSS/C_{14}TAB mixture at 10^{-3} M, a special phenomenon is observed: Shortly after film formation, a black dot occurs in the film. However, the spreading of the domain is very slow and does not finish even after hours. Only when the applied pressure is increased, the spreading continues and the film ruptures soon after the pressure increase. It seems that the difference in interfacial tension inside and outside the domain is so small, that the two film thicknesses are in equilibrium and the spreading is not favored, so that the film is trapped in a metastable state. Apparently, the surface layer is close to maximal coverage, since a NBF with a small diameter is stabilised but not crystalline enough to stabilise the further growing film.

Furthermore, aggregates between NaSS and $C_{14}TAB$ are formed at the highest investigated concentration. These are observed in Fig. 7.16 (right) as a bright layer on the NBF, but are not visible in the bulk solution. The aggregates are not fixed at the surface as found for polyelectrolyte/surfactant gels,[39] but move freely on the surface. However, the aggregates seem to disturb the crystalline surface structure of the NBF such that the film ruptures soon after the transition. Like in case of polyelectrolytes, the threshold of the aggregate formation seems to be a surfactant chain length of 14 alkyl groups, since the interactions between the monomer and $C_{12}TAB$ are not so strong that visible complexes are formed. Even at monomer concentrations that are one order of magnitude higher, no aggregates can be observed, which can easily be explained by a reduced hydrophobic interaction between the compounds.

In case of $C_{10}SO_3$, no NBF transitions are observed around the cmc. In contrast to the mixtures with NaSS, the stability of the foam film strongly increased above the nominal IEP. Furthermore, the foam films containing $C_{10}SO_3$ get thinner due to the increasing ionic strength in the system. This is supported by the surface potentials, which are reduced to around 30 mV at 10^{-3} M $C_{10}SO_3$. These observations are in good agreement with findings of other surfactant foam films in the literature.[16,97] Very stable films with stabilities up to 6800 Pa are formed in this concentration regime but no stratification occurs until a concentration that is around one order of magnitude higher than the cmc. Eventually, at a concentration of 10^{-1} M $C_{10}SO_3$, several steps in thickness can be observed that are assumed to be due to a layering of micelles in the system. After the stratification a film with a thickness of 10 nm is formed. The last transition seems to be a NBF transition, but unfortunately, at this particular thickness of 10 nm it is hard to say, if really a NBF is formed or if it is rather a very thin CBF. However, the surface coverage seems to be to low or the repulsion of the interfaces not strong enough to stabilise the film, so that it ruptures at rather low disjoining pressures.

In summary, the monomer NaSS has the character of a cosurfactant and a organic salt, depending on its concentration in the mixtures. At low concentrations the cosurfactant character dominates and it has the ability to stabilise foam films. However, it can not stabilize foam films on its own and can therefore not be compared to a real surfactant. Especially in the high concentration regime, the effect of the NaSS and $C_{10}SO_3$ differs significantly. In case of NaSS, no stratification occurs, but a CBF-NBF transition. The deviation from the surfactant properties is mostly due to the hydrophobic part of the monomer. It is indeed hydrophobic enough to be accessible to hydrophobic interactions with the surfactant, but the hydrophobic part is too short, so that the chains of the two compounds in the system are not so compatible to form strong complexes. Furthermore, at concentrations close to the cmc, the salt effect starts to dominate hydrophobic properties of the monomer, which leads to the formation of a NBF.

7.3.3 Comparison between monomer and polymer

In this section, the effect of the respective monomer on foam film properties is compared to the effect of the polyelectrolyte discussed in chapter 4 and 5. In case of AMPS/PAMPS with $C_{14}TAB$, both differences and similarities are observed. As in films with the corresponding polymer, stable foam films are formed at low concentrations. However, these films are much more stable than those of the pure surfactant, whereas in case of PAMPS, the stability was reduced below the IEP. Additionally, the surface potentials in this regime are much higher for the AMPS systems, which also explains the high stability of the foam films. Close to the nominal IEP, a complete destabilisation of the foam film occurs, which resembles the PAMPS/$C_{14}TAB$

7.3 Discussion

system. However, above the IEP, a jump in film stability is observed, which is in contrast to the steady increase in film stability that occurs when polyelectrolyte is added to the system. Furthermore, when the AMPS concentration is increased, the stability is reduced until a complete destabilisation of the foam films at 10^{-2} M, which is contradictory to the results of the polymer.

In case of C_{12}TAB, the effect on foam films is again different. First of all, AMPS, like PAMPS, has the ability to stabilise foam films, so that stable foam films are formed even at low AMPS concentrations around 10^{-5} M. Yet in contrast to films with PAMPS, the increase of the amount of monomer has almost no influence on the foam film stability until a concentration of 5×10^{-4} M. The surface potentials and the surface tension in this concentration regime are more or less constant, which supports the assumption that the addition of monomer has only a minor effect on the surface coverage. Furthermore, in case of PAMPS/C_{12}TAB the destabilisation of foam films was not very distinct, but for the AMPS system, no destabilisation or stability minimum can be observed at all. This implies that the mechanism that leads to the destabilisation effect is different, when the surfactant chain is prolonged. At higher concentrations, the foam film stability decreases and a complete destabilisation occurs, which is again contradictory to the films with PAMPS. In this concentration regime, the stability of both AMPS/C_nTAB systems coincide, even though surface tension measurement show distinct differences in surface coverage.

Additionally, the surface tension measurements reveal that the characteristics of the surface layers are very different. In case of polyelectrolyte/surfactant mixtures, the surface tension was strongly reduced below the IEP and then abruptly changed at the IEP. In contrast to this, the surface tension of the monomer mixtures remains constant at a value close to that of the pure surfactant and is only decreased, when a strong charge screening occurs at high monomer concentrations.

In general, one can say that, even though in the concentration range below the nominal IEP the foam film properties are quite similar to the polyelectrolyte/surfactant system, the effect of the AMPS can be better explained by the influence of the salt character. This is especially obvious in the surface tension measurements, where no hydrophobic complexes are observed at the interface and all curves show a continuous behaviour. The characteristics that distinguish AMPS from the simple salt originate in the volume of the ion and its slight hydrophobicity.

The influence of the monomer NaSS is very different from the one of the corresponding polyelectrolyte PSS. In case of PSS, the interactions between the polyelectrolyte and C_{14}TAB are so strong that aggregates are formed already at quite low polymer concentrations, so that it is impossible to investigate the foam film of this mixture. When C_{12}TAB is used in the system, the polyelectrolyte stabilises the foam films. This is also the case for NaSS at low concentrations but the effect is much stronger. When the concentration is increased no stable foam films can be formed in a broad concentration regime. For the polymer a minimum in film stability is observed close to the IEP, but no complete destabilisation occurs. This leads to the conclusion, that the stabilisation process is different in the two systems. At high concentrations, the foam films with PSS are very stable and CBFs are formed. In contrast to that, NaSS does not form stable CBFs above the IEP but NBF are observed in the concentration regime above 10^{-2} M. In this concentration regime, the monomer shows rather the characteristics of a salt.

The surface tension shows significant differences between the two systems as well. The isotherms of NaSS/C_nTAB mixtures show the typical shape of a surfactant with a continuous decrease and a plateau at the cmc. This implies that micelles are present in the solution. On the other hand, PSS/C_nTAB systems show the characteristic non-monotonous behaviour of the polyelectrolyte/surfactant mixtures discussed in previous chapters.

In summary, the properties of NaSS are better described by a cosurfactant than by the polymer. Due to the hydrophobicity of the monomer, it can adsorb at the surface over a large concentration regime in the presence of surfactant and the foam film are similar to catanionic mixtures presented above. However, in the high concentration regime acts more like an organic salt.

7.4 Conclusion

The results of the present study show that the influence of the monomer differs significantly from the effect of the polymer, even though the foam film stabilities seem similar at first sight. In case of AMPS, the addition of the monomer has similar effects as the addition of a simple salt like NaCl. Foam films below the IEP are stabilised due to its ability to reduce the repulsion between the surfactant head groups and the resulting increased adsorption of the surfactant. Above the IEP, the increased ionic strength in the film enhances the ion condensation at the film surfaces and decreases the repulsion of the opposing film interfaces. This leads to a reduction of the film thickness and film stability and finally to the destabilisation of the foam films in the high concentration regime.

In contrast to that, NaSS behaves like a cosurfactant and stabilises foam films after the addition of very low amounts. However, in the concentration regime around the IEP, no stable films can be formed even though the surface coverage of the film interfaces is quite high. This is assumed to be due to a low surface elasticity when the monomer with the short hydrophobic tail is adsorbed. At concentrations around the cmc of the system, CBF-NBF transitions can be observed. A transition to a NBF is typical for systems, where high amounts of salt are added, which suggests, that in this concentration regime the salt character of the monomer starts to affect the foam films.

In summary, the functionality of the molecules strongly affects the interactions between the compounds in the mixtures, so that it is crucial, if a monomer or a polymer is added to the surfactant solution. Furthermore, besides the size of the molecule, the hydrophilic/hydrophobic balance in the monomer unit plays an important role in foam film stabilisation.

8 Effect of the polyelectrolyte chain length

Abstract

This chapter deals with the influence of the polyelectrolyte chain length on foam film stabilities of oppositely charged polyelectrolyte/surfactant mixtures. In chapter 7, the influence of a long polyelectrolyte chain and the effect of the monomer have been investigated. The differences in the findings suggest that the polyelectrolyte chain length affects the foam film properties. In this study, rather short polyelectrolytes with 60 and 20 repeat units, respectively, have been investigated to gain information about the transition from monomer to polymer effect.

The results show that above a certain chain length, the molecular weight of the polyelectrolyte has only a minor influence on the foam film stability and the surface tension. However, the use of a polymer with 20 monomer units reveals that beside the described effect of the polymer and the monomer, also an oligomer effect can be observed. In this case, the adsorption process and the foam films stabilities are considerably different to the former findings, due to the short polyelectrolyte chain.

8.1 Introduction

Polyelectrolyte/surfactant mixtures are often used in industrial applications like personal care products and cleaning agents. Besides the rheological properties of the bulk solution, one can also change the adsorption and the foam film characteristics by choosing different polyelectrolyte and surfactant types. In the case of oppositely charged polyelectrolyte/surfactant mixtures, strong electrostatic interactions occur which lead to the formation of complexes between the two compounds. This is both, energetically and entropically favorable, since the repulsion of the surfactant headgroups at the interface is reduced and the entropy of the system is increased due to the release of the counterions.[27]

In chapters 4 and 5, foam films of cationic surfactant and anionic polyelectrolytes around the IEP have been investigated. In the case of $C_{12}TAB$ and PSS with a chain length of > 300 monomer units, the addition of polyelectrolyte to the surfactant solution leads to the formation of stable foam films and a minimum in film stability at a concentration very close to the nominal IEP. From surface tension and elasticity measurements it follows that there is a reduction in surface charge at low PSS concentrations but no charge reversal takes place at the interface. In a second approach, the polymer was exchanged by the monomer, which is in the case of PSS, NaSS. As discussed in chapter 7, the foam film properties of the mixtures with the monomer are very different from those with the polymer. Due to the hydrophobic parts of the molecule, the monomer resembles a cosurfactant and can be integrated into the surface layer at all investigated concentrations and has also the ability to form micelles in presence of another surfactant.

From these results, the question arises, how the foam films are affected, when instead of the long polyelectrolyte or the monomer, different PSS with short polyelectrolyte chains are added to the mixture. Can a transition between the monomer behaviour and the effect of the polymer be observed?

In a study by Monteux et al.[33] it was shown that a polymer with at least 20 monomer units is needed to get the effect of the polymer in terms of surface tension isotherms. In the case of an oligomer with less monomer segments, the adsorption at the air water interface differed strongly from that of longer polymers. For macromolecules with more than 20 monomer units, no significant dependence on molecular weight has been found. In general, the polyelectrolyte adsorption at the interface and, therefore, the surface tension should be constant at the same monomer unit concentration, when the polymer adsorbs flat at the interface.[126] In the case of coil adsorption or when the adsorption rate of the polymer endgroup differs from that of the middle segments, changes should be visible in the surface tension.[127]

The molecular weight dependence in thin liquid films in the semidilute concentration regime, is already well established.[128–132] In this concentration range, the molecular weight has no influence on the oscillatory force in the system. However, when the longer polyelectrolytes are exchanged by a polymer with 20 monomer units, the scaling behaviour changes from a power law that scales with $c^{-1/2}$ to one with an exponent of $-1/3$.[104] This implies that the polyelectrolytes do not overlap to a network-like structure in the film bulk but arrange as spherical coils. In more detail, the transition between the dilute regime and the critical overlap concentration (c^*) is shifted to higher concentrations, when a shorter polyelectrolyte is used.[128]

Nevertheless, there are some properties in the film bulk that do have a molecular weight dependence, e.g. the osmotic pressure. The shorter the polyelectrolyte chain, the higher is the osmotic pressure in the film at the same monomer concentration.[130] Altogether, this suggests that shorter polyelectrolytes do affect the foam films in the investigated concentration regime.

To investigate this influence of the polyelectrolyte chain length on the foam film stability, PSS samples of different molecular weights have been studied and compared to former measurements of PSS/C_{12}TAB mixtures with a chain length of about 340 monomer units and to foam films where the corresponding monomer is added. The IEP of the system is determined by the used surfactant concentration of 10^{-4} M. Besides the disjoining pressure isotherms of the foam films of the mixtures, also surface tension and partly elasticity measurements have been performed and are presented in the following section.

8.2 Results

PSS of two different molecular weights, namely 20 monomer units (PSS20) and 60 monomer units (PSS60) were studied.

8.2.1 PSS60/C_{12}TAB

Disjoining pressure isotherms at different PSS60 concentrations are shown in Fig. 8.1. The surfactant concentration is fixed at 10^{-4} M, which defines the nominal IEP at 10^{-4} M. The foam films are characterised by the film thickness, the stability, i.e. the maximum pressure that can be applied to the film before the film ruptures (Π_{max}), the slope of the isotherms that depends on the ionic strength of the sample solution and the surface potential which has been simulated by using the PB program version 2.2.1,[12] written by Per Linse. This program solves the nonlinear Poisson-Boltzmann equation under the assumption of constant potential.

Fig. 8.1a depicts isotherms below the IEP, while Fig. 8.1b shows the isotherms at and above the IEP. The isotherms up to a concentration of 10^{-4} M PSS60 almost coincide on one curve an dif-

8.2 Results

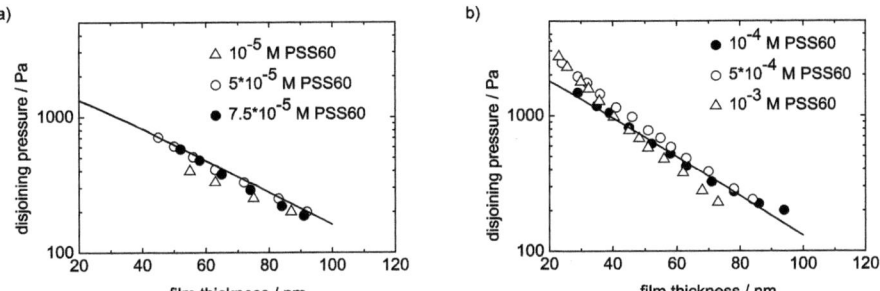

Figure 8.1: *Disjoining pressure isotherms of PSS60/C_{12}TAB solutions with varied polyelectrolyte concentration; a) below the IEP; b) at and above the IEP; the solid lines correspond to the simulation of the isotherm with constant potential; for the sake of clarity, only some of the simulated isotherms are shown in the graph.*

fer only in slope and stability. The film thickness at a disjoining pressure of 200 Pa is 91-95 nm, which is in the same range as the thicknesses in other investigated polyelectrolyte/surfactant systems in this concentration regime. At higher polymer concentrations the equilibrium thickness is reduced to 83 and 72 nm, at 5×10^{-4} M and 10^{-3} M PSS60, respectively due to a stronger screening of the electrostatic double-layer. This is in contrast to other studied polyelectrolyte/surfactant systems, where no screening could be observed at high polyelectrolyte concentrations.

Furthermore, the slope of the isotherms is affected by the polyelectrolyte concentration. This effect is related to the ionic strength in the system and the isotherms get steeper with increasing polymer concentration. At concentrations below the nominal IEP, the effect is not so pronounced, but at concentrations $\geq 5 \times 10^{-4}$ M PSS60, a strong increase in the slope of the isotherms is observed.

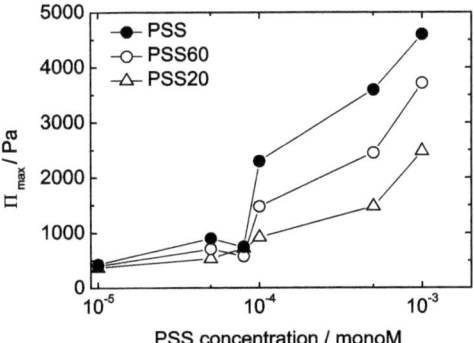

Figure 8.2: *Stability of PSS/C_{12}TAB films with different PSS chain lengths; maximum disjoining pressure Π_{max} before film rupture versus polyelectrolyte concentration.*

The addition of 10^{-5} M PSS60 leads to an increase in stability compared to the pure surfactant

PSS60 conc. [monoM]	Ψ_0 [mV]	κ^{-1} [nm]	I [M]
1×10^{-5}	52	29.9	8.0×10^{-5}
5×10^{-5}	60	28.5	9.0×10^{-5}
7.5×10^{-5}	58	26.0	1.0×10^{-4}
1×10^{-4}	65	24.5	1.2×10^{-4}
5×10^{-4}	78	22.1	1.7×10^{-4}
1×10^{-3}	65	14.2	3.0×10^{-4}

Table 8.1: *Summary of the surface potentials Ψ_0 from the simulation of the disjoining pressure isotherms of PSS60/C_{12}TAB films; the Debye length κ^{-1} is calculated from a fit of the experimental data with a exponential decay function of first order, the ionic strength I is derived from the simulation of the surface potential.*

system (cf. chapter 5) to a maximum pressure of 430 Pa (see Fig. 8.2). A polymer concentration of 5×10^{-5} M further increases the stability to 750 Pa. However, at 7.5×10^{-5} M PSS60, Π_{max} is reduced and a minimum in film stability of 600 Pa is observed. Further addition of PSS60 is again accompanied by an increasing film stability, which is far beyond the stability that could be reached below the IEP. Nevertheless, the film stabilities are with 1500 Pa at 10^{-4} M and 4500 Pa at 10^{-3} M PSS60, respectively, almost 1000 Pa lower than in case of PSS with a chain length of 340 monomer units.

The shape of the stability curve is also reflected in the surface potentials. The surface potential of the foam films first increases from 52 to 60 mV when the polyelectrolyte concentration is changed from 10^{-5} M to 5×10^{-5} M PSS60, shows a minimum of 58 mV at 7.5×10^{-5} M and then further increases to 78 mV upon addition of 5×10^{-4} M PSS60. However, even though the stability of the foam film further increases with an increase of the polyelectrolyte concentration to 10^{-3} M, the surface potential decreases to 65 mV due to the charge screening in the system. The results of the surface potential are summarised in Table 8.1.

Surface tension measurements of the corresponding polyelectrolyte/surfactant solutions are shown in Fig 8.3. The surface tension isotherm of PSS60 corresponds to the typical shape of polyelectrolyte/surfactant isotherms found in Ref.:[90,107] The addition of low amounts of polymer (10^{-5} M) lead to a strong reduction in surface tension compared to the pure surfactant solution. Further increase of PSS60 concentration reduces the surface tension to a minimum of 58 mN/m at the nominal IEP. Comparing this to the surface tension of PSS/C_{12}TAB solutions at the respective concentration reveals, that the shorter polyelectrolyte chain decreases the surface tension 1-2 mN/m more than the polyelectrolyte with the higher molecular weight. At PSS60 concentrations $\geq 10^{-4}$ M, a sudden increase in surface tension to 70 mN/m is observed, indicating a release of surface complexes and leaving only a more or less pure surfactant layer at the interface. Above the nominal IEP of the system, both PSS and PSS60 surface tension curves almost coincide. This suggests that the mechanism that leads to the depletion of the surface in the same for both polyelectrolytes.

8.2.2 PSS20/C_{12}TAB

As described in section 8.1, the differences of the disjoining pressure isotherms of PSS60/C_{12}TAB are only minor compared to the PSS with 340 monomer units and occur mainly at in the high polyelectrolyte concentration range. To get a deeper understanding of the influence of the chain length on film stability, a shorter polyelectrolyte is studied as well. According to Monteux *et al.*,[33] a polyelectrolyte chain length of at least 20 monomer units is needed to reach polymer

8.2 Results

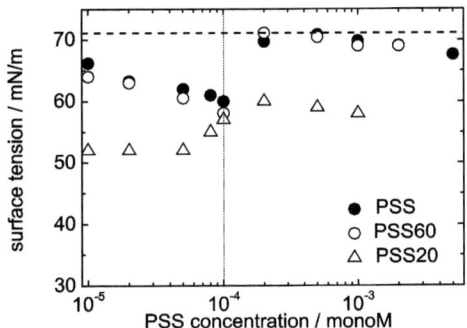

Figure 8.3: *Surface tension of PSS/C_{12}TAB solutions with fixed surfactant (10^{-4} M) and variable PSS concentration; the dashed line corresponds to the surface tension of the pure surfactant; the vertical line illustrates the nominal IEP of the system.*

character, so that a PSS with this chain length has been chosen.

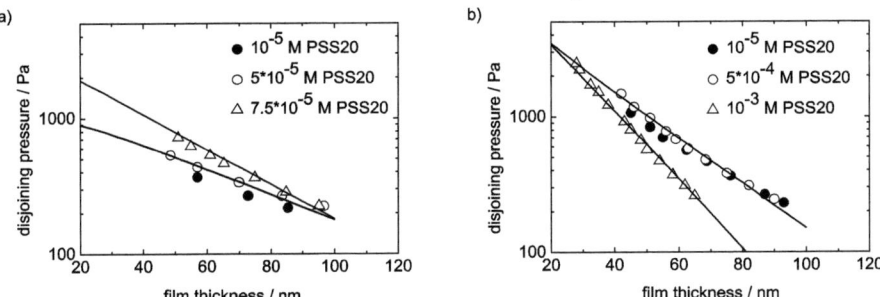

Figure 8.4: *Disjoining pressure isotherms of PSS20/C_{12}TAB solutions with varied polyelectrolyte concentration; a) below the IEP; b) at and above the IEP; the solid lines correspond to the simulation of the isotherm with constant potential; for the sake of clarity, only some of the simulated isotherms are shown in the graph.*

Disjoining pressure isotherms of the PSS20/C_{12}TAB mixtures are shown in Fig. 8.4 and divided into isotherms below the nominal IEP in Fig. 8.4a and at and above the IEP, respectively, in Fig. 8.4b. Up to a concentration of 5×10^{-4} M PSS20, all foam films start with a equilibrium thickness of 95-98 nm at 220 Pa. Only at 10^{-3} M, the charge screening due to the higher ionic strength sets in and reduces the film thickness to 64 nm at this point. The effect of the screening is even more pronounced for this polyelectrolyte than in the case of PSS60. This trend is also found for the slope of the disjoining pressure isotherms. The Debye lengths calculated from the fit of the disjoining pressure isotherms with an exponential function of first order (*cf.* Table 8.2), are much lower than those of the PSS60 system. The increased ionic strength in the system leads to steeper isotherms, and the differences are already distinct below the nominal IEP.

PSS20 conc. [monoM]	Ψ_0 [mV]	κ^{-1} [nm]	I [M]
5×10^{-5}	60	29.0	9.0×10^{-5}
7.5×10^{-5}	67	27.1	1.2×10^{-4}
1×10^{-4}	70	22.5	1.4×10^{-4}
5×10^{-4}	75	18.9	2.0×10^{-4}
1×10^{-3}	60	13.7	3.5×10^{-4}

Table 8.2: *Summary of the surface potentials Ψ_0 from the simulation of the disjoining pressure isotherms of $PSS20/C_{12}TAB$ films; the Debye length κ^{-1} is calculated from a fit of the experimental data with a exponential decay function of first order, the ionic strength I is derived from the simulation of the surface potential.*

With increasing polyelectrolyte concentration, the stability of the foam films continuously increases (*cf.* Fig. 8.2). This is different to all other polyelectrolyte/surfactant mixtures that showed a destabilisation or minimum stability at 7.5×10^{-5} M. Nevertheless, the stability increases from 380 Pa at 10^{-5} M to 580 Pa at 5×10^{-5} M and 880 Pa at 7.5×10^{-5} M PSS20. This difference of 300 Pa is within the accuracy of the measurement, so that this effect can be taken into account. Comparing all systems, one clearly sees that the film stabilities at 7.5×10^{-5} M coincide on one disjoining pressure and differ only below and above this point. When the PSS20 concentration is further increased, the film stability increases from 1050 Pa at 10^{-4} M to 3000 Pa at 10^{-3} M. Additionally, it is shown that the trend at polyelectrolyte concentrations $\geq 10^{-4}$ M is the same, but the absolute film stabilities are further reduced compared to long polyelectrolyte chains. The surface potentials of the isotherms summarised in Fig. 8.2, support this picture. The foam film with the lowest polyelectrolyte concentrations could not be simulated, since only 3 data points are available. However, the surface potential increases from 60 mV at 5×10^{-5} M to 75 mV at 5×10^{-4} M. Only at 10^{-3} M PSS20, the surface potential is reduced to 60 mV due to strong charge screening in the system. Besides the differences at the particular concentration of 7.5×10^{-5} M, all potentials are in the same range than those of $PSS60/C_{12}TAB$.

Surface tension isotherms of $PSS20/C_{12}TAB$ mixtures are shown in Fig. 8.3. Comparing measurements of these mixtures with the ones of solutions containing polyelectrolytes with a higher molecular weight, the effect on surface tension is completely different. Firstly, the surface tension is already lowered to 52 mN/m at a concentration of 10^{-5} M, which is 12 mN/m lower than that of PSS60. Additionally, the surface tension stays constant until a PSS20 concentration of 5×10^{-5} M and than steadily increases. This increase starts below the nominal IEP of the system, *i.e.* at lower concentrations than for the other PSS systems. The sudden increase of the surface tension detected for both PSS and PSS60 at a polyelectrolyte concentration above the IEP is not observed in the case of PSS20. Furthermore, the surface tension goes not back to the value of the pure surfactant, but reaches a maximum of 60 mN/m, indicating that the polyelectrolyte/surfactant complexes remain at the interface. However, the surface tension slightly decreases at polyelectrolyte concentrations higher than 2×10^{-4} M, which is which is consistent with $PSS/C_{12}TAB$ mixtures.

Both the foam film and the surface tension measurements of $PSS20/C_{12}TAB$ show significant differences to the results of polyelectrolyte/surfactant mixtures with a longer polymer chain, so that additional experiments are required. To get a deeper insight into the surface properties of the system, surface elasticity measurements have been performed at different frequencies between 0.005 and 0.1 Hz (*cf.* Fig. 8.5). Again, the results differ strongly from those of $PSS/C_{12}TAB$ in chapter 5. At low PSS20 concentrations, the surface elasticity of 88 mN/m

8.3 Discussion

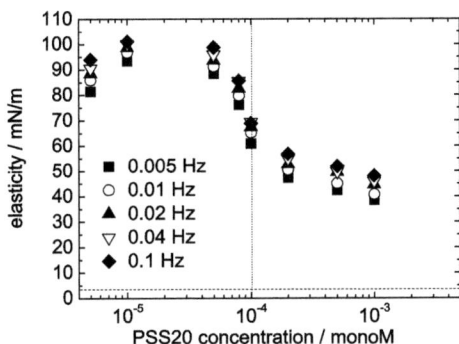

Figure 8.5: *Surface elasticity of PSS20/C_{12}TAB solutions with fixed surfactant (10^{-4} M) and variable PSS concentration at different frequencies; the dashed line corresponds to the elasticity of the pure surfactant; the vertical line illustrates the nominal IEP of the system.*

±6 is highly elevated compared to that of the pure surfactant solution. The elasticities of the different oscillation frequencies are distributed over a range of 12-15 mN/m, which is constant over the whole concentration range. At a concentration of 10^{-5} M PSS20, the maximum elasticity of about 100 mN/m is reached and stays almost constant, when the polymer concentration is further increased to 5×10^{-5} M. Close to the IEP of the system, the surface elasticity starts to decrease again. However, not a sudden reduction of surface elasticity like in case of PSS/C_{12}TAB is observed, but rather a steady and continuous decrease. At 10^{-4} M PSS20 for example, the surface elasticity is lowered to 65 mN/m and it is reduced to 46 mN/m at 10^{-3} M, the highest investigated concentration, which is not even close to the elasticity of the pure surfactant. At this concentration, the decrease seems to level off, which coincides with surface tension measurements at the respective concentration, where a slight decrease is detected. Altogether, the results suggest that the addition of PSS20 to C_{12}TAB solutions leads to a completely different adsorption of the polyelectrolyte/surfactant complexes than in case of longer polyelectrolyte chains.

8.3 Discussion

The focus of this study was the investigation of the influence of the polyelectrolyte chain length on the foam film stability. In general, the results of the foam film stabilities suggest, that the effect of the polyelectrolyte with 60 and 20 monomer units resembles more that of the long polymer chain than that of the monomer. In the latter case, the foam film stabilities and the surface characteristics are completely different.

The foam film stabilities reveal that the addition of polyelectrolyte leads to the formation of stable foam films with C_{12}TAB. The surfactant has, due to the short hydrophobic tail a low adsorption rate to the air/water interface, which results in a low surface charge and, in turn, in unstable C_{12}TAB foam films.[17] However, when PSS is added, regardless which of the three types, stable films are formed in the whole investigated concentration regime. For the two long chain polyelectrolytes, the foam film stability and the surface tension isotherms have the characteristic shape, discussed in chapter 5. The foam film stability first increases when polymer is added and a minimum in stability occurs close to the nominal IEP of the system. Once the IEP is crossed, very stable foam films are formed. Although the foam film stabilities below the IEP are very similar, the films of the PSS mixtures are in average 1000 Pa more stable than those with PSS60.

As mentioned above, the surface tension isotherms of the surfactant solutions with PSS and PSS60 are very similar. Below the IEP, very hydrophobic complexes between C_{12}TAB and the polyelectrolyte are formed, which adsorb at the interface due to the strong electrostatic interaction between the two compounds. Since more surfactant molecules than polyelectrolyte segments are present in the solution, all charges of the polymer chain are complexed by surfactant molecules, which expose their hydrophobic alkyl chains to the surrounding. The release of the counterions further increases the entropy of the system, which also enhances the complexation between the compounds. Additionally, the hydrophobic backbone of the polyelectrolyte can be adsorbed at the surface. Altogether, this leads to a strong decrease in surface tension already at quite low polyelectrolyte concentrations. With further increase of the PSS concentration, the surface tension is reduced until a minimum at the nominal IEP. The surface tension measurements in this concentration regime show, that PSS60 adsorbs slightly stronger at the surface than the longer polyelectrolyte. This indicates that the length of the polyelectrolyte chain indeed has an influence on the adsorption. As described earlier, a change in surface tension gives a hint that the polyelectrolyte does not adsorb flat to the interface.[126] In that case, no change in surface tension would be visible since the monomer concentration and therefore the amount of adsorbed segments would remain constant. On the other hand, when the polyelectrolyte is adsorbed in a coiled conformation, the amount of adsorbed coils increases, because a higher amount of polymer chains is available at the same monomer concentration when a shorter polyelectrolyte is added.

At concentrations above the nominal IEP of 10^{-4} M, a significant change of the surface coverage is observed. A strong increase in the surface tension gives evidence that the polyelectrolyte/surfactant complexes are released from the surface and a more or less pure surfactant layer is left at the interface. Above the IEP, an excess of polyelectrolyte segments is present in the system. Not every charged monomer unit can be complexed by an surfactant molecule which increases the charge on one polyelectrolyte chain. This in turn makes the complexes more hydrophilic and less surface-active. Since no surface-active complexes remain at the surface in this concentration regime, the surface tension behaviour of the two polyelectrolytes is the same above the IEP.

The adsorption process of PSS20/C_{12}TAB complexes at the surface is more complex. Not only the absolute surface tension is reduced, but the shape of the isotherm is changed as well. Below the IEP, a strong decrease in surface tension is observed. Furthermore, it remains constant until a concentration of 5×10^{-5} M and is not continuously reduced until a minimum at the nominal IEP as it is the case for the two longer PSS. On the contrary, a steady increase in surface tension takes place, that starts at 7.5×10^{-5} M PSS20 and continues until 2×10^{-4} M. These findings are supported by the elasticity measurements, where a slow decrease is observed in the same concentration regime. The difference to the systems with PSS and PSS60 is manifested not only in the contrast between stepwise change in surface tension in the former case instead of the sudden release of surface complexes in the latter, but also the striking fact that no complete release of the polyelectrolyte/surfactant complexes is observed. The surface tension increases to a level that corresponds to the minimum of the mixtures with the longer polyelectrolytes.

For the explanation of these significant differences in the adsorption, the changes in the system have to be considered. Firstly, the use of short polyelectrolyte chains does not change the

8.3 Discussion

concentration of monomer units but that of the polyelectrolyte endgroups. When the endgroups and the monomer segments in the polymer chain have a different tendency to adsorb at the surface, this can lead to a significant change in the adsorption process.[127] In the case of PSS60, the endgroup concentration is increased by a factor of 6 and in the PSS20 system is increased already 17 × which could lead to strong changes. However, the hydrophobicity difference between -CH_2- and -CH_3 is only minor and can not explain the dramatic changes observed for the PSS20 system.

The low surface tension and the high surface elasticity that are observed at concentrations between 10^{-5} M and 5×10^{-5} M in the PSS20/C_{12}TAB mixture gives further evidence on the structure of the surface layer. Especially the high surface elasticity of almost 100 mN/m implies, that a rigid, interconnected surface layer is formed.[133] PSS/C_{12}TAB can interact electrostatically as well as hydrophobically due to the oppositely charged monomer units and the hydrophobic backbone of the polymer.[107] Therefore, the surfactant can connect the polymer chains at the surface and form a network-like surface layer. The strong reduction in surface tension also leads to the assumption that the polyelectrolytes adsorb in a more extended fashion to the surface, which leads to a strong decrease in surface tension even at very low polyelectrolyte concentrations. Further addition of polyelectrolyte to the mixture results in an increase in the surface tension, while the elasticity is reduced. This indicates that the interconnected surface network disappears and the conformation of the adsorbed polyelectrolytes changes. Since more polyelectrolyte chains are available in this concentration regime and short chains have a higher diffusion coefficient,[126] the polymer chains are assumed to adsorb in a more coiled fashion.

At concentrations above the nominal IEP, the surface tension is reduced by about 10 mN/m compared to the PSS and PSS60 system. In the latter case, this is explained by a lower amount of surfactant molecules that are complexed with the polyelectrolyte, which makes the aggregate less surface-active. However, since the monomer unit/surfactant ratio has not been changed, the reduced surface tension has to be originated in a different adsorption process to the surface. It is suggested that a cooperative binding process occurs at the interface when a short polyelectrolyte chain adsorbs to the surface, since the short polymer is more flexible and can diffuse faster. Once a polymer is connected to the interface, it can easily complex with more surfactant molecules that are in close proximity of the polyelectrolyte chain. This would lead to a non-uniform distribution of the surfactant molecules on the polymer and could explain the special surface tension characteristics of PSS20. The slowly decreasing surface elasticity in the concentration range above 2×10^{-4} M additionally indicates a formation of multilayers at the surface.

Besides the number of polyelectrolyte chains in the system and the surface coverage, also the dissociation degree of the polyelectrolyte's charged groups is affected. This already is visible in the disjoining pressure isotherms of PSS60 at a concentration of 10^{-3} M where the surface potential is reduced to 65 mV compared to 85 mV in case of PSS.[107] For PSS20 at the same concentration, the potential is further reduced to 60 mV. Additionally, the foam films are formed with lower films thicknesses at the described concentration, compared to the mixtures with a long polyelectrolyte, namely 72 nm for PSS60 and 65 nm for PSS20, respectively. The ionic strengths calculated from the exponential fit of the disjoining pressure isotherms, shows that the ionic strength in the system is increased from 3.3×10^{-4} M for PSS to 5.0×10^{-4} M for PSS20 at a concentration of 10^{-3} M. The high ionic strength does not agree with the Manning concept of polyelectrolyte condensation which was valid in all investigated systems with longer polyelectrolyte chains. Both the high ionic strength and the low film thicknesses are more the characteristics of a added salt, which leads to the conclusion that the polyelectrolyte character is reduced when a shorter polyelectrolyte is used.

Below the IEP of the mixture, the foam film stabilities of all three systems are rather similar. Especially at the lowest investigated concentration of 10^{-5} M PSS, all three foam films have the same film stability. This and the fact that the surface potentials of the foam films are similar imply that the molecular weight of the polyelectrolyte plays only a minor role, although the characterisation of the surface layer shows already significant differences. However, above the nominal IEP, the foam film stabilities differ strongly between the polyelectrolyte/surfactant systems. In this concentration regime, the foam film stabilities follow the trend PSS > PSS60 > PSS20. If this is due to the increasing polymer chain concentration and their higher mobility in the film core[92] or because of the lower surface potential for shorter polymer chains in the high concentration regime, remains speculative.

The foam film stabilities reveal another peculiarity of PSS20 compared to all other investigated polymer/surfactant systems. For the other mixtures, a complete destabilisation or a minimum in foam film stability has been observed close to the nominal IEP. In contrast to that, PSS20 does not show any stability minimum at the polymer concentration of 7.5×10^{-5} M, but a continuous increase in film stability. However, the foam film stabilities in this system are in principal quite low compared to the other mixtures, due to the higher polyelectrolyte chain concentration and the higher mobility of the macromolecules in the film core. The results show, that the film stabilities of all three systems almost coincide on one point at this special concentration. Furthermore, the surface tensions are in the same range at this point so one could assume that different polymer chain length indeed do change the adsorption and foam film characteristics, but at the IEP of the system, this does not play a role. On the contrary, at this certain point, only the polyelectrolyte/surfactant ratio in the system seems to be important.

Further evidence, that PSS20 has different properties than the longer polyelectrolytes show the following experiments by Üzüm et al.:[104] When the PSS20 concentration is further increased, stratification can be observed (data not shown), which is a hint for a polymer network and, therefore, characteristic for polyelectrolytes. However, the investigation of PSS20 solution in thin films between two solids surfaces has shown that the structuring of the polymer chains is different with the short polyelectrolyte with 20 monomer units. Polyelectrolytes in the semidilute concentration regime form a network of overlapping polymer chains both in thin liquid films and in the bulk solution. When the two opposing film interfaces are approached, an oscillating force is measured. The distance between the oscillation maxima is concentration dependent and scales with a power law of $c^{-1/2}$.[52] The exponent of $-1/2$ is characteristic for polymer networks. On the other hand, in the case of spherical particles and micelles, the power law scales with $c^{-1/3}$. In the case of a short polyelectrolyte chain, the overlap concentration c^* is shifted to higher concentrations,[128] which is manifested in a scaling law of $c^{-1/3}$ for PSS20. This indicates that short polymers do not overlap, but arrange as single coils in the film core.

In summary, the results reveal that in broad range the molecular weight plays only a minor role, but for smaller polymers with just a few monomer units, the influence changes significantly.

8.4 Conclusion

In this study, TFPB, surface tension, and elasticity measurements have been performed to investigate the effect of the polyelectrolyte chain length on foam film stability and adsorption process. Compared to the PSS with 340 monomer units, almost no differences can be observed, when PSS with a chain length of 60 repeat units is added. Only the increased number of chains and the resulting higher fluctuations in the film core lead to the formation of less stable films compared to the longer polyelectrolyte. Yet, the characteristic shape of the foam film stabilities and the surface tension isotherms remain the same.

8.4 Conclusion

However, the presented results show that the use of an oligomer with 20 repeat units indeed has an influence on both, adsorption and foam film stabilities. The higher concentration of polyelectrolyte chains in the films results in an increased osmotic pressure and a faster diffusion/mobility of the polyelectrolytes in the foam film. Furthermore, the short polymer chains adsorb flat at the film interface at low polyelectrolyte concentrations and change their conformation to a more coiled structure, when the concentration is increased. In contrast to the other mixtures, no release of surface complexes occurs above the IEP and the adsorbed polymer coils remain at the surface. The described changes have an impact on the foam films and lead to lower film stabilities. Just at the IEP, where a minimum stability has been observed for all polyelectrolyte/surfactant systems, no destabilisation occurred. Instead, the stabilities of all three systems coincide on point, which results in an continuous increase in film stability for the $PSS20/C_{12}TAB$. This implies that the foam film stabilities in this concentration regime are mainly dominated by polyelectrolyte/surfactant concentration ratio in the film.

Altogether, the results presented in this chapter show that, besides the polymer and the monomer effect discussed in previous chapters, an oligomer effect can be found for polyelectrolytes with 20 repeat units.

9 Dynamics of polymer chains in thin films

Abstract

In the following chapter, the dynamics of polymer chains in thin films are investigated by using polyelectrolytes labeled with the fluorescent dye rhodamine B. In the first section, the behaviour of polyelectrolytes in the foam film during the stratification process is investigated. The results indicate that the polyelectrolyte layers are not expelled from the network in the film core, which was the hitherto assumption, but collapse onto the remaining layers. However, further experiments have to be conducted, to explain the discrepancies between the presented results and those of former studies.

In the second part of the chapter, the diffusion of polyelectrolytes in the film core is studied and compared to that in the bulk solution. For this purpose, fluorescence correlation spectroscopy (FCS) and surface force apparatus (SFA) measurements have been performed. It was shown that the confinement in the foam film indeed influences the diffusion of the macromolecules. Compared to the bulk solution, the diffusion is slowed down, when the polyelectrolyte solution is confined between the two interfaces of the film due to the similar dimensions of the film thickness and the polymer coil. However, when the film thickness is further reduced, the polymer chain diffusion is accelerated again.

9.1 Introduction

In the previous chapters, the influence of polyelectrolyte/surfactant mixtures on foam film stability was investigated. However, besides being a good method to get information about foam film stabilities, foam films can also be used to study colloidal particles or macromolecules in spatial confinement. Confinement effects are likely to occur, when molecules are brought into an environment with dimensions that are similar to their own size. Since this is the case in thin liquid film with a thickness around 100 nm, they present a good model system of a slit pore.[57]

In the following, results on the structuring and the diffusion of polyelectrolytes in thin films are presented. For these investigations, polyelectrolytes are labeled with a fluorescent dye, which makes them accessible to single molecule-tracking or the measurement of fluorescence intensity fluctuations. It is important to use covalently bound dye molecules, since otherwise, the dye could be expelled from the thin liquid film and no information about the polyelectrolyte chains could be given.

In this study, rhodamine B was used, which is a a strong fluorescent dye with an emission maximum at 585 nm and a stable molecule that is not so sensitive to quenching and bleaching as fluorescein, for example.

9.2 Fluorescence spectroscopy on foam films

The conformation of polyelectrolyte chains strongly depends on the concentration. For example, in the semi-dilute regime, above a certain threshold concentration, polyelectrolytes start to overlap and form network like structures in bulk solution. The mesh size ξ of this network depends on the monomer concentration and scales with $c^{-1/2}$. When the macromolecules are entrapped between two surfaces and the interfaces approach each other, this gives rise to an oscillatory force. In foam films, between two air/water interfaces, the oscillations manifest themselves as a stepwise thinning of the film, also referred to as stratification. The step size is independent of the surfactant type, which influences only the initial film thickness. This implies that the stratification arises only from the polyelectrolyte structuring and not from interactions between polyelectrolyte and surfactant.[15] The force oscillations can also be observed between two solid surfaces, as colloidal probe-AFM measurements show[52,61] and are therefore entirely independent of nature of the surface.

The origin of this oscillatory force is controversially discussed in literature. Initially, Milling et al.[134] proposed a layering of coiled polyelectrolyte chains in the film. However, this supposes a dependency of the step size on the chain length, which could not be confirmed in studies with polyelectrolyte of different molecular weight.[53] Since the period between the oscillations corresponds to the mesh size of the polymer network in the bulk, it is now assumed that a similar transient network is formed in confinement.[52,135] Therefore, the steps in film thickness are supposed to be associated with the number of meshes that fit the distance between the interfaces. When the pressure on the interfaces is increased, the polyelectrolyte network rearranges such that a network with one mesh less is formed, which is visible in the step in thickness. In an alternative model, a layer-wise arrangement of the polyelectrolyte chains with a predominant alignment parallel to the confining surfaces[51] is proposed. This suggestion is in good agreement with theoretical predictions,[132] but fails to explain the connection between the step size of the stratification and the mesh size in bulk solution. So far, the arrangement of the polyelectrolytes in confinement remains an open question.

Solving this problem is related to the question, what happens during the stratification process. In a former work of Toca-Herrera et al.[49] on fluorescein-labeled poly(allylamine hydrochlorid) (PAH), the changes in fluorescence intensity before and after the step in film thickness has been investigated. For that reason, a foam film was formed in a TFPB and the fluorescence intensity was monitored while the disjoining pressure in the film was increased. In this study, it was shown that the measured emission intensity decreased during the stratification process, which was explained by a layer-wise expulsion of the labeled polyelectrolytes from the film.

On the other hand, in a study on the eximer formation in foam films from pyrene labeled poly(acrylic acid) (PAA), the total intensity of the fluorescence emission remained constant during the stratification process. This, in turn, implies that the total number of polyelectrolyte chains in the film is constant and that the network is rearranged such that instead of being expelled, one layer of the polyelectrolytes collapses onto the remaining layers. To clarify these contradictory results, the fluorescence of foam films containing rhodamine B-labeled PSS (RITC-PSS) is investigated.

9.2 Fluorescence spectroscopy on foam films

9.2.1 Additional experimental details

Fluorescence-TFPB

For these experiments, a modified TFPB is used that is connected to a fluorescence microscope and a spectrometer. The fluorescence microscope (Eclipse 600 FN, Nikon) is equipped with a filter block for rhodamine B excitation (AHF-Analysetechnik, Tübingen, Germany). In this block, an excitation filter, transmitting at 543 ± 11 nm, is combined with a dichroic mirror (562 nm) and an emission band pass filter (590 ± 20 nm). The sample is illuminated by a mercury short arc lamp. Additionally, a further block provides the possibility of observing the film with white light to determine the film thickness. The fluorescence signal is analysed by a spectrometer (Triax 180, Jobin Yvon-Spex Instruments S.a., Cedex) that is coupled to the microscope and the spectra are recorded with an accumulation time of 10 × 1 s. The images of the film are captured with a fluorescence sensitive high-resolution CCD camera (Hr 1UV, Proxitronic, Bensheim, Germany) that can be connected to the microscope instead of the spectrometer.

9.2.2 Results and discussion

Fluorescence-TFPB is used to investigate the behaviour of the polyelectrolytes during the stratification process in foam films from mixtures of $C_{12}TAB$ and RITC-labeled PSS. In the semidilute concentration regime a stepwise transition to thinner foam films is observed. Depending on the polyelectrolyte concentration one or more steps are induced. In this experiment, a concentration of 10^{-2} M is used, where of 90% is unlabeled polyelectrolyte and 10% of the polymer chains carry a RITC- label. The degree of labeling is about 1 dye per 100 monomer units, which leads to a total rhodamine B concentration of 10^{-5} M. To stabilise the foam films, a concentration of 10^{-4} M $C_{12}TAB$ is used. As shown in Fig. 9.1, this mixture leads to foam films in which one stratification step is observed. The step width is 19 nm from a film thickness of 59 to 40 nm at a disjoining pressure of 150 Pa. After the transition, the film is stable up to 900 Pa before it ruptures.

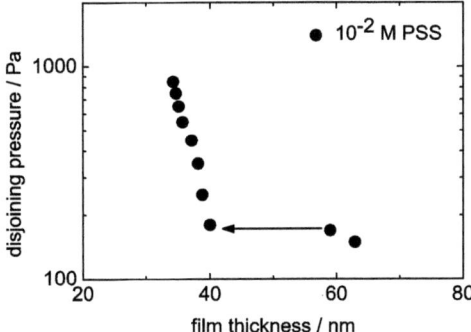

Figure 9.1: *Disjoining pressure isotherm of $PSS/C_{12}TAB$ in the semidilute concentration regime; the step in thickness is illustrated by the arrow.*

For the observation of the fluorescence intensity, the applied pressure is kept constant for 20

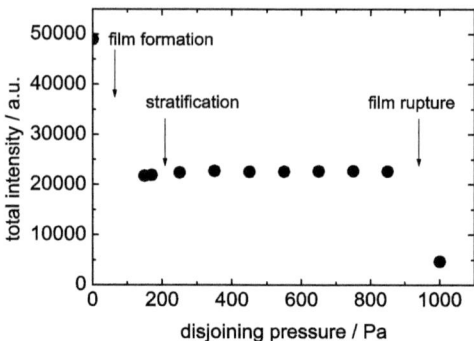

Figure 9.2: *Representation of the change in fluorescence intensity versus disjoining pressure.*

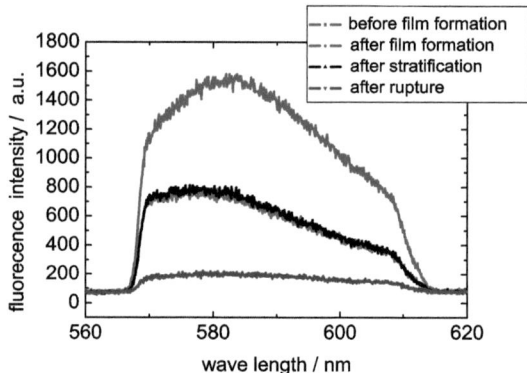

Figure 9.3: *Fluorescence spectra of the different states of the foam film.*

min and then a fluorescence spectrum is measured. To quantify the intensity, the area under the spectrum is calculated. In Figs. 9.2 and 9.3, the results of these measurements are shown. In Fig. 9.2, the fluorescence intensity is plotted versus the disjoining pressure. Before the film formation, the fluorescence of the droplet in the film holder is measured. After the film formation, the intensity is reduced to less than a half. However, during the stratification process, the intensity of the emitted light does not change significantly, according to the calculated intensities, the increase is about 1 %. In Fig. 9.3, the fluorescence spectra at the different film states are depicted.

Since the peak maximum is not so distinct and the emission spectra seem to be very broad, they have been compared to the ones of a RITC-PSS bulk solution, recorded with a commercially available fluorescence spectrometer. The spectra of a concentration range between 5×10^{-5} M and 10^{-3} M are shown in Fig. 9.4 and follow the same trends as the spectra of the foam film. With increasing dilution, the maximum disappears more and more and the emission spectrum gets very flat. According to these results, the existence of stray light that disturbs the measurements can be excluded. However, due to the band-pass filter, only a part of the

9.2 Fluorescence spectroscopy on foam films

whole fluorescent peak is recorded, so that the shape of recorded spectra seems to be very unusual.

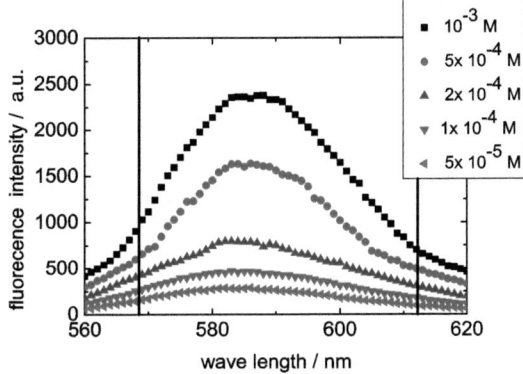

Figure 9.4: *Fluorescence spectra of different bulk concentrations of RITC-PSS; the solid lines indicate the cut-off wave length of the emission band pass filter*

The stratification of the foam film can also be observed via a fluorescent-sensitive CCD camera. The images of the transition are shown in Fig. 9.5, where the brighter spots correspond to the thin film areas. The thinning of the film starts at several areas of the film simultaneously. The spots grow and coalesce, thereby spreading over the whole film. However, on the fluorescent images, the intensity increase seems to be much stronger than that detected by the spectrometer.

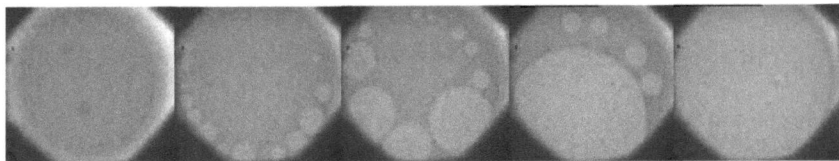

Figure 9.5: *Images of the changes of fluorescent intensity during the stratification process.*

The presented results show that the polyelectrolyte concentration remains constant in the film core during the stratification process. So far is was assumed that single layers of the polyelectrolyte network in the film core are expelled and the amount of polyelectrolyte is reduced. However, since the constant intensity in the measurements indicates a constant polyelectrolyte concentration in the film, this assumption can not hold anymore. It is rather proposed, that the network is rearranged during the transition to a thinner film. One possible explanation would be that one layer collapses onto another layer in the film, and, this way, all polyelectrolytes are integrated into the remaining network.

Our results are supported by the findings of Rapoport et al.,[57] who also found a constant total fluorescence intensity after the stratification step in foam film with pyrene-labeled PAA. However, the opposite effect has been observed for fluorescein-labeld $PAH/C_{12}G_2$ mixtures.

To clarify these contradictions and to understand the behaviour of the polyelectrolyte network in the film core when the applied pressure is increased, the influence of different parameters has to be investigated. Firstly, the effect of the fluorescent dye that is used to label the polyelectrolytes is considered. For example fluorescein that is used in the latter case, is known to be sensitive to bleaching and to self-quenching. The reduction of the fluorescence intensity was interpreted as a decrease of polymer concentration in the film, but it could be also due to a self-quenching process when one polyelectrolyte layer collapses onto the remaining network. In this case, the polyelectrolytes are in close proximity and are accessible to the self-quenching.

Another important parameter is the surfactant type and the interactions between the two compounds in the film. The type of surfactant has no influence on the step width of the stratification process and only affects the initial film thickness.[15] However, it does have an impact on the spreading velocity of the domain on the film. When polyelectrolyte and surfactant interact at the interface the domain growth in much slower than in the case of polymer/surfactant combinations, which do not form complexes. This is explained by polyelectrolyte chains that dangle into the film bulk, thereby slowing down the stratification process.[62,63] These dangling chains could also influence the stratification process by hindering the polyelectrolytes to be expelled from the film core. To study the effect of the surfactant during the stratification, it would be interesting to compare it to measurements without any surfactant. Since foam films without surfactant are not stable, theses experiments should be performed on a solid support. For this reason, one could combine fluorescence measurements with a CP-AFM, where aqueous films are confined between two solid surfaces.

So far, the results have been interpreted as constant fluorescent intensities. However, the measurements show a slight increase in intensity. Why the fluorescence seems to increase after the stratification remains unclear. It is rather unlikely, that the concentration of polyelectrolyte in the film increases during the rearrangement of the network in the film core.

In summary, the findings indicate, that the behaviour of the polyelectrolyte is different from the proposed mechanism of the expulsion of polyelectrolyte layers from the film. However, to clarify the mechanism and to understand the influence of parameters like surface properties of the fluorescent labels, further experiments need to be conducted.

9.2.3 Conclusion

The presented results on the changes on the fluorescence intensity during the stratification process of foam films lead to a surprising conclusion: The intensity of the emitted light remains almost constant or rather increases after the stratification step, which is in contrast to the findings of a former study. Furthermore, it is contradictory to the former assumptions that a polyelectrolyte layer is expelled from the film during the stratification. The results indicate that instead, the polyelectrolyte layer collapses onto the remaining layer in the film core, so that the amount of polyelectrolyte is constant during the experiment. However, further studies need to be conducted to elucidate the contradictions and their correlation to the type of the fluorescent dye, the surface charge *etc.*

9.3 Diffusion of polyelectrolytes in thin films

As mentioned above, thin films also correspond to a slit pore geometry and can be used as a tool to study the effect of confinement on polyelectrolytes. The thickness of the films is several tens of nm, so that the dimension of the film is in the same range as that of the macromolecules. The confining walls of the film affect the polymer such that the macromolecules get a new equilibrium conformation that differs from the bulk state, which also changes the dynamic properties of the polymer as, for example, diffusion.[136] The effect on diffusion and molecular conformation

9.3 Diffusion of polyelectrolytes in thin films

are of great interest in molecular biology applications such as nanofluidics. Nanofluidic devices contain nanoslits and are used for analytical processes like the separation of biomacromolecules by electrophoresis or HPLC.[136,137]

Besides the confinement, the diffusion is influenced by a large variety of parameters like the viscosity of the solution, salt concentration, molecular size, and degree of ionisation of the macromolecule.[138,139] A well established method to study these effects in bulk solutions, is the fluorescence correlation spectroscopy (FCS).[140,141] In this approach, the diffusion coefficient is calculated from the diffusion time through a defined volume in the solution. However, to investigate the diffusion of macromolecules in confinement, it is better to couple a fluorescence microscope to a surface force apparatus (SFA). A droplet of the aqueous solution containing the polyelectrolyte is placed between two glass slides and the thickness is adjusted. To follow the diffusion of the polymer in both cases, the macromolecule has to be labeled with a fluorescent dye. The polyelectrolyte of choice is poly(allylamin hydrochlorid) (PAH), which is covalently bound to fluorescing molecules in a simple procedure. Furthermore, the degree of ionisation can be easily adjusted by controlling the pH in the system.[142] The critical assumption of this setup is that the attachment of a small dye molecular does not substantially change the properties of the polymer chain. However, this requirement seems to be fulfilled.[139]

Since the experiments in the thin liquid films are conducted between two solid interfaces, no surfactant is needed to stabilise the film. This has the advantage, that no interactions between polyelectrolytes and surfactants can alter the diffusion process. Furthermore, to avoid the adsorption of the positively charged polyelectrolytes to the negatively charged glass walls, the surface has to be modified. The silanisation of the glass slides results in a surface that carries positive charges at the desired pH of 5.5.

In the following, the diffusion of polyelectrolytes in confinement is investigated and compared to the behaviour in the bulk solution. For this reason the positively charged polyelectrolyte PAH was labeled with rhodamine B, which was also used in the first section of this chapter. In a first approach, the bulk diffusion was studied by FCS experiments and the effect of a confinement to 50 μm was tested. To gain more information, how the confinement to several tens of nm affect the diffusion, single molecule-tracking experiments have been performed in a SFA and the film thickness has been varied between 40 and 150 nm.

9.3.1 Additional experimental details

Fluorescence correlation spectroscopy (FCS)

Fluorescence correlation spectroscopy (FCS) experiments were performed in cooperation with Christine Papadakis from the TU München. FCS was developed in the 1970s and is a sensitive method to study kinetic processes through statistical analysis of equilibrium fluctuations. A fluorescent dye is coupled to the system of interest, so that spontaneous fluctuations in the system's state generate variations in fluorescence intensity. The autocorrelation function of the fluorescence emission fluctuations carries information on the characteristic time scales of the system.[140,141] For example, the typical decay time of the autocorrelation function is related to the diffusion time of the particles through the detection volume. The fluorescence intensity, $I(t)$, emitted by the fluorescent molecules in the observation volume is measured and autocorrelated with the following equation.

$$G(\tau) - 1 = \frac{\langle \delta I(t) \delta I(t+\tau) \rangle}{\langle I(t) \rangle^2} \tag{9.1}$$

Figure 9.6: *Scheme of polystyrene spacers between two glass slides.*

where $\delta I(t)$ is the instantaneous deviation of the measured intensity from its time average, $\langle I(t) \rangle$. For freely diffusing, uniformly fluorescing, and monodispers particles, the correlation function can be fitted with the following expression:

$$G(t) = \frac{1}{N}\left(1+\frac{t}{\tau_D}\right)^{-1}\left(1+\frac{t}{\omega^2 \tau_D}\right)^{-1/2}\left(1+\frac{p}{1-p}\exp\left(-\frac{t}{\tau_T}\right)\right) \tag{9.2}$$

where N is the average number of molecules in the sampling volume, p is the fraction of the dye molecule in the triplet state, τ_D is the diffusion time and τ_T is the triplet time.[140,143] To calculate the diffusing coefficient from the fitted diffusion time, the setup is calibrated before starting the experiment. This is usually done by measuring the diffusion time of molecules with a known diffusion coefficient, in this case rhodamine 6G (Sigma-Aldrich, $D = 2.8 \times 10^{-10}$ m^2/s).

To study the effect of the viscosity on the diffusion times and coefficients of the polyelectrolytes, it was varied by adding different amounts of glycerol to the solution, namely 10 and 50 w/w %. The diffusions coefficients are then calculated by the following equation.

$$D_{PAH} = D_{Rh}^{Gl/H2O}\frac{\tau_{Rh}^{Gl/H2O}}{\tau_{PAH}^{Gl/H2O}} = D_{Rh}^{H2O}\frac{\eta_{H2O}}{\eta_{Gl/H2O}}\frac{\tau_{Rh}^{Gl/H2O}}{\tau_{PAH}^{Gl/H2O}} \tag{9.3}$$

with D being the diffusion coefficient, Gl/H$_2$O the corresponding water/glycerol mixture, τ_x the respective diffusion time and η the viscosity of the solutions.

The FCS measurements were carried out at a LSM 510- Confocor 2 from Carl Zeiss Jena GmbH (Jena, Germany). The sample was placed in a sample holder and illuminated with a HeNe-laser with excitation wave length of 543 nm. The fluorescent fluctuations were recorded for 10 × 10 s, with a long-pass filter of 585 nm. To investigate the effect of confinement on the diffusion of the polyelectrolytes, the diffusion coefficients of the bulk solutions have been compared to those in thin liquid films. A simple way to create thin films is to trap the polyelectrolyte solution between two glass slides using polystyrene beads with a defined diameter in the micrometer range as spacers. (*cf.* Fig. 9.6).[144]

To prevent the adsorption of the positively charged polyelectrolytes at the negatively charged glass substrates, it is important to modify the surfaces. Therefore, the glass slides were silanised by dipping them into an aminosilane solution prior to the experiment.

Surface force apparatus (SFA)

The surface force apparatus (SFA) was first set up in the 1970s and has been improved and modified since then.[145,146] It is usually taken to investigate interactions in thin liquid films between two solid surfaces such as structural forces, drainage or phase transitions. However, these experiments are restricted to the measurement of mean values of physical properties and are therefore unable to reveal molecular motions, as the diffusion constant or molecular mobility in confined space. Hence, the combination of the SFA technique with a video microscope for the

9.3 Diffusion of polyelectrolytes in thin films

tracking of single fluorescent dye molecules provides more detailed information on the diffusion and mobility of single molecules.

The experiments were performed in cooperation with Frank Cichos and Martin Pumpa from the Universität Leipzig. The home-build set up used in this study consists of a highly sensitive wide field fluorescence microscope, with the SFA placed beneath. The whole apparatus is mounted on vibration isolated table and the sample is illuminated by an frequency doubled Nd:YAG laser with a wave length of 532 nm. The fluorescence is than collected by a CCD camera for further evaluation. The excitation light is separated from the luminescence by a dichroic beam splitter and blocked by a long-pass detection filter.

The fluorescent probe is placed beneath two quartz glass plates that are glued to an aluminium frame. The upper surface is fixed to the SFA, while the lower surface is mounted on a piezo scanner and a xyz- translation stage, so that it can be moved in all three dimensions.[147] To determine the distance between the two surfaces, gold nanospheres with a diameter of 40, 60, 80, 100, and 150 nm are used as spacers. Due to the monodispersity of the particles, a precise determination of the film thickness is possible.

For the determination of the diffusion coefficients, videos of 1000 frames with a exposure time of 20 ms and a frame rate of 50 Hz were recorded. The evaluated area was 150×150 Pixel, which corresponds to 21×21 μm. The step size distribution is determined be comparing two consecutive frames and connecting bright areas with a distance of less than 10 pixels with trajectories. The diffusion coefficient can then be calculated by the following equations.[148,149]

$$D = \frac{\sigma_x^2 + \sigma_y^2}{4\tau} \tag{9.4}$$

where σ_i is the standard deviation of the step size in x and y direction.

$$D^{obs}/D^{real} = 1 - \frac{t_{exp}}{3t_{frame}} \tag{9.5}$$

with t_{exp} being the exposure time and t_{frame} the time between the frames.

PAH-Labeling

The cationic polyelectrolyte poly(allylamin hydrochlorid) (PAH, $M_W = 65000$) was purchased from Sigma-Aldrich (Steinheim, Germany) and used as received. For the fluorescent labeling, 25 mg PAH were dissolved in 2 mL carbonate buffer (pH = 9.5). 200 μL of a 1 mg/mL rhodamine B isothiocyanate solution (RITC) (Fluka, Steinheim, Germany) were added to the polyelectrolyte solution and stirred for 2 h. After the reaction time, the labeled PAH fraction was separated from the unbound dye via size exclusion chromatography with a column from Amersham Bioscience (Freiburg, Germany). The resulting rhodamine-labeled polyelectrolyte was freeze-dried and dissolved to the desired concentration.

Modification of the glass surfaces

To avoid the adsorption of positively charged polyelectrolyte to the negatively charged glass substrate, the surfaces were modified such that the resulting surface charge is positive. For this purpose, cover slips were cleaned in a 1:1 H_2O_2:H_2SO_4 solution for 20 min and then carefully rinsed with Milli Q water. Afterwards, the glass was dipped in a 1 % aqueous solution of 3-aminopropyltrimethoxysilane for 10 min. During this step, the silane is covalently bound to the Si-OH groups at the glass surface, which results in a positively charged surface. After the

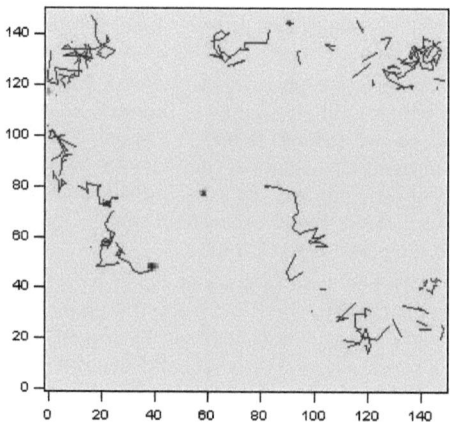

Figure 9.7: *Snap shot of the molecule tracking process; the trajectories correspond to the diffusing polyelectrolytes, the bold points to adsorbed polymer chains and the small dots to free rhodamine B molecules; done in cooperation with the University Leipzig.*

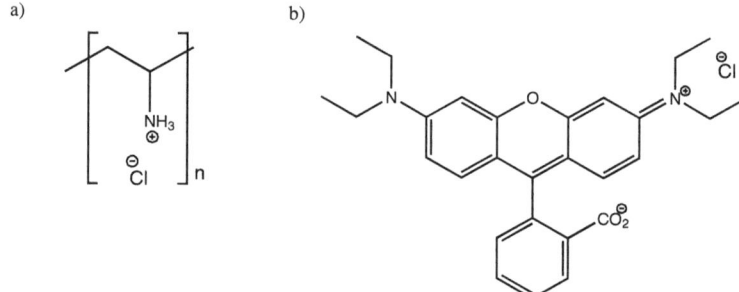

Figure 9.8: *Chemical structure of a) PAH and b) the fluorescent dye rhodamine B.*

dipping, the slips are carefully rinsed with EtOH and Milli Q water and placed in an oven for 10 min at 110 °C.[150]

9.3.2 Results and discussion

To determine the effect of confinement on diffusion, first the diffusion of the polyelectrolytes in the bulk solution has to be investigated. The method of choice is the FCS which allows the determination of the diffusion time and coefficient.

In Fig. 9.9, the autocorrelation functions of RITC-PAH in different H_2O/glycerol mixtures are shown. Glycerol increases the viscosity of the solution, which slows down the diffusion of the polyelectrolyte. A slower diffusion is desired, since in single molecule tracking, slower particles can be followed more easily. In an aqueous solution the diffusion coefficient is 1.1×10^{-9} m^2/s. The addition of 10 % glycerol to the polyelectrolyte solution has almost no effect on the diffusion, in this low concentration regime, the influence of glycerol on the viscosity can

9.3 Diffusion of polyelectrolytes in thin films

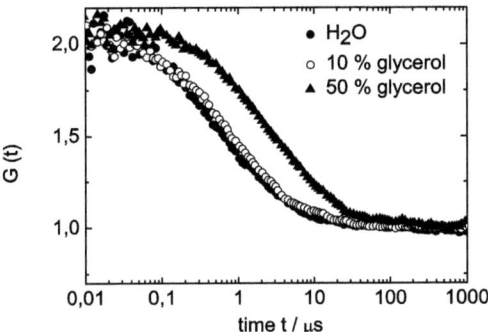

Figure 9.9: *Normalised FCS autocorrelation functions of fluorescent-labeled PAH in water/glycerol mixtures; done in cooperation with the TU München.*

be neglected. However, when 50 % glycerol are added, the effect on viscosity is much more pronounced and the diffusion coefficient is reduced to 3.6×10^{-10} m^2/s.

To determine diffusion of the polyelectrolytes in thin films in the μm range, the liquid is confined between two glass slides and polystyrene beads are used as spacers. The thickness of the film is determined by a z-scan through the film and was found to be 50 ± 5 μm. The position of the sample volume can also be adjusted via the z-scan. This way, both the diffusion in the film core and close to the interface can be studied. However, in this thickness range, the influence of confinement is only minor, so that the diffusion coefficients are not affected at all. Furthermore, no difference between the polymer diffusion in the film core and in close proximity to the confining walls can be detected.

Since further confinement can not be reached with a commercial FCS without any adaption single molecule tacking experiments in confinement have been performed on a SFA. In these experiments, only polyelectrolyte solutions with 50 % glycerole are investigated, since only diffusion coefficients in this range are accessible to the method. The film thickness is again adjusted by using spacers between the two glass slides, but in this case gold nanoparticles with a much smaller diameter are used, namely 150, 100, 80, 60, and 40 nm. After acquisition of the videos, the diffusion coefficients are determined by the method described above. In Table 9.1 the results of the diffusing polyelectrolytes in confinement are summarised. The diffusion coeffcients reveal that the confinement indeed affects the diffusion of the polyelectrolyte. However, the evolution of the diffusion is not monotonous. The confinement to a film thickness of 150 nm, the diffusion coefficient is reduced to 8.1×10^{-12} m^2/s which is almost two orders of magnitude lower compared to the bulk solution. Further confinement leads again to an increase of the diffusion coefficient. In the thinnest film with a thickness of 40 nm, the polyelectrolytes are accelerated to 1.0×10^{-10} m^2/s, which is the same order of magnitude than that of the bulk solution.

In the bulk solution, the polyelectrolytes have a coiled conformation. Depending on the degree of charge, the hydrodynamic radius of the polyelectrolyte chain is different. For polymers with a high degree of charge, the radius of the coil increases due to the electrostatic repulsion between the charged monomer units, and according to this, polyelectrolyte chains with a lower degree of charge are smaller.[142] In case of PAH, the degree of charge can be easily adjusted by

film thickness [nm]	diffusion coefficient [m^2/s]
150	8.1×10^{-12}
100	8.9×10^{-12}
80	1.4×10^{-11}
60	1.6×10^{-11}
40	1.0×10^{-10}

Table 9.1: *Diffusion coefficients of RITC-PAH in a 50 % gylcerol mixture in confinement between 150 and 40 nm.*

changing the pH of the system. At the used pH of 5.5, the polyelectrolyte is highly charged which increases the hydrodynamic radius.

When the molecule is brought into a confinement of several tens of nm, the dimensions of the polymer are in the same range as the surrounding geometry. This has an effect on the equilibrium state of the coil and changes its conformation.[136] The change in conformation would explain the initial reduction in diffusion coefficient upon confinement. It is supposed that the structure of the polymer changes from a spherical coil to a more stretched conformation. Long and needle-like structures are more sensitive to the confinement,[151] which would explain the observed effect.

The non-monotonous behaviour of the diffusion coefficient is rather surprising. However, there are theoretical considerations that explain the increase of the diffusion coefficient for spherical particles with further confinement of the film.[152] When the polyelectrolytes diffuse in the bulk solution the surrounding of the macromolecules is rather rough due to neighbouring polymer chains. The neighbouring macromolecules hinder the diffusion, since the friction between the coils is rather large. Yet, when the polyelectrolytes are close to the borders of the film, the walls are smooth compared to the surrounding of the coils in bulk solution. This reduces the friction between the macromolecule and the wall and the diffusion process is accelerated. With decreasing film thickness, the influence of the smooth wall gains importance, since more molecules are in close proximity to the confining walls. Therefore, the diffusion coefficient increases with increasing confinement.

9.3.3 Conclusion

In this section, the influence of the viscosity and the confinement on polyelectrolytes is investigated. The increase in viscosity alters the diffusion of the polymer chains, but the effect is only visible at glycerol concentrations of 50 %. The addition of this amount of glycerole reduces the diffusion coefficient by about one order of magnitude. Furthermore, the presented results show that the confinement influences the diffusion of the polyelectrolyte as well. First, the confinement to 150 nm decreases the diffusion coefficient compared to the bulk, but further confinement again accelerates the diffusion. It is assumed that the confinement changes the equilibrium conformation of the polyelectrolyte chain to a more elongated coil, which, in turn, induces the reduction of the diffusion coefficient. However, further confinement leads to an acceleration of the diffusion. In thinner films, the influence of the confining wall is enhanced; the smooth walls reduce the friction of the polymers compared to the neighbouring polymer chains in the volume phase. The combination of these effects leads to the non-monotonous change of the diffusion coefficient.

10 Conclusion and Outlook

General conclusions

The main focus of this work is the effect of oppositely charged polyelectrolyte/surfactant mixtures on foam films around the IEP of the system. Polyelectrolyte/surfactant mixtures were the subject of many studies in the last years, but most of them consider the bulk properties or the surface characterisation and if foam films are discussed, mainly the semidilute concentration regime of the polyelectrolyte is studied. The articles on foam film with polyelectrolyte/surfactant mixtures in the dilute concentration regime are few.

In this thesis, mainly mixtures of positively charged surfactant and negatively charged polyelectrolytes in the dilute concentration regime around the IEP of the mixture were investigated. The foam films have been studied with respect to different aspects:

(i) the effect of the polymer concentration

(ii) the influence of the size of the macromolecule and the surfactant

(iii) the effect of the hydrophilic/hydrophobic balance of the molecules

(iv) the influence of small additives like a salt

(v) the dynamics of polyelectrolytes in the film core

The presented results show that the foam films are very sensitive to the polymer/surfactant concentration ratio and to the type and hydrophobicity of the two components. To study foam films around the IEP of the systems, the excess charge of the polyelectrolyte/surfactant complexes was varied by using a fixed surfactant concentration and a variable amount of polyelectrolyte. In most investigated systems, a minimum in stability could be observed close to the nominal IEP. However, the exact shape of the foam stability curve was strongly affected by the used surfactant. In case of $C_{14}TAB$ and highly charged polyelectrolyte, the foam film stability was reduced towards the IEP and no stable foam films could be formed at the point of equal charges. Once this point was crossed, the stability of the films increased again with increasing polyelectrolyte concentration. On the other hand, in case of $C_{12}TAB$, the foam film stabilities first increased upon polyelectrolyte addition and only a minimum stability was observed at the IEP. Above the IEP, very stable foam films were observed. In general, no stable foam films can be obtained from pure $C_{12}TAB$ solutions. Nevertheless, the addition of all types of negatively charged additives as hydrophobic and hydrophilic polyelectrolytes, with a long or short chain, as well as the monomer or the salt that were investigated, always led to the formation of stable foam films. Depending on the systems, the concentration regime in which the initial stabilisation appeared was different.

The general properties of foam films formed from polyelectrolyte/surfactant mixtures were very similar throughout all systems. A reduction of foam film stability was detected slightly below the nominal IEP of the system and very stable foam films were found in the concentration regime above the IEP. However, the surface characterisation of the air/water interface revealed that this phenomenon is not due to a charge reversal at the interface. The shape of the stability curve is qualitatively the same for different surface coverages and is dependent on the polyelectrolyte/surfactant concentration ratio. Below the IEP, hydrophobic polyelectrolyte/surfactant complexes adsorb at the surface, but due to the low amount of unbound surfactant molecules in the system, the foam films are not very stable. Furthermore, above the IEP, most of the surface-active complexes are released from the interface and only a more or less pure surfactant layer is left at the surface. Nevertheless, in this concentration regime, the most stable foam films are found. The qualitative influence of the polyelectrolyte hydrophobicity on foam film stability was only minor, at least in the case of 100 % charged polyelectrolytes. The addition of more hydrophobic polyelectrolytes only affected the absolute film stabilities, but not the shape of the stability curve.

In all investigated polyelectrolyte/surfactant mixtures, only two exceptions were found, where the surface tension differed from the described behaviour. A copolymer with only 60 % charged monomer units was used to proof that the destabilisation effect at the IEP was independent from the position of the IEP. The adsorption of the polyelectrolyte at the interface was changed due to the uncharged parts of the macromolecule that increased the hydrophobicty of the polymer. Furthermore, the surface characterisation revealed that the polyelectrolyte is accessible to more than electrostatic interactions between the charged groups and hydrophobic interactions between the hydrophobic tail of the surfactant and the polymer backbone. The charged head group of the surfactant can also interact with the dipole moment of the uncharged units of the polyelectrolyte. This way, the positive charges of $C_{14}TAB$ were shielded far below the IEP, which led to the destabilisation of the foam film at quite low polyelectrolyte concentrations. In a similar set of experiments, where the IEP was shifted by reducing the surfactant concentration, it was shown that a certain amount of unbound surfactant molecules is needed to stabilise the foam film, since a broad range of unstable films around the IEP appeared in these mixtures as well.

The second exception was the addition of polyelectrolytes with only a few repeat units instead of the longer polymers with 340 monomer units used in a former study. When a polymer with 60 monomer units was used, the surface properties and the foam film stability were only slightly affected. Only the film stabilities were shifted to slightly lower values due to the higher amount of polymer chains and the resulting higher fluctuations in the film. However, the addition of polyelectrolytes with only 20 monomer units had a strong influence on both the adsorption and the foam film stabilities. The surface tension was reduced much stronger, both below and above the IEP. This was due to the fact that the short polyelectrolyte adsorbed in a more extended way and a cooperative binding process occurred, once the polyelectrolyte was connected to the surface. Even above the IEP, polyelectrolyte/surfactant complexes remained adsorbed at the surface. The film stabilities were affected by the short polymer chains as well. They were further reduced (lower Π_{max}) compared to the longer polyelectrolytes. However, the stability minimum at the IEP was not very pronounced. On the contrary, at this particular point, the surface tension and the foam film stabilities of all three investigated polyelectrolyte chain lengths coincided in one point. It is suggested that at this particular concentration, only the polyelectrolyte/surfactant ratio is important. Altogether this leads to the conclusion that in a broad range of molecular weight, the influence of chain length of the foam film stability and the surface coverage is only minor. However, in the case of oligomers with only a few repeat units, the effect on stability and adsorption is more pronounced.

In general, the addition of small molecules like the monomer or salt led to different foam film properties in comparison to that of the polyelectrolytes. The influence of the hydrophilic/hydrophobic balance plays an important role concerning foam film stability and surface composition. When the hydrophilic monomer AMPS is added to the solution, the effect is comparable to the addition of a monovalent salt. The surface tension is continuously reduced due to the screened repulsion between the surfactant headgroups. Below the IEP, stable foam films were formed, but with increasing ionic strength in the mixture, both the film thickness and the stability were reduced. One exception of the continuous behaviour of the foam film stability was found at the IEP of the system. At this particular point, a destabilisation of the foam films occurred as it was observed for polyelectrolyte/surfactant mixtures. Again, the destabilisation was more pronounced for $C_{14}TAB$ than for $C_{12}TAB$ where no stability reduction could be observed. In case of NaSS, the more hydrophobic monomer, the molecule was integrated into the surface layer and led to a strong reduction in surface tension. At low concentrations the monomer has the ability to stabilise the foam films, which gives evidence for a characterisation as a cosurfactant. However, in the high concentration regime, a CBF- NBF transition was observed, indicating that in this range the monomer resembles an (organic) salt.

In the last chapter of the thesis, the dynamics of polymer chains in thin films were investigated. Firstly, the effect on the polymer network in the semidilute concentration regime during the stratification was investigated by means of fluorescence measurements. The results indicated that polyelectrolytes are not expelled from the film core during the transition, which was the former working hypotheses, but rather collapsed onto the remaining network. However, contradictions to former results remain to be solved. In the second part, the diffusion process in confinement is studied with single molecule-tracking. The combination of different effects leads to a non-monotonous evolution of the diffusion coefficient. First, the confinement changes the conformation of the polyelectrolyte chain, which results in a decrease of diffusion coefficient compared to the bulk behaviour. In thinner films, the diffusion coefficient increases again because the reduced friction of the smooth walls compared to the neighbouring polyelectrolytes in the bulk phase gains more importance.

Future prospectives

During the work on this project and the unraveling of some of the initial questions, new issues developed which could be further explored. In the following, some possible directions for future research will be presented.

The investigation of foam films from oppositely charged polyelectrolyte/surfactant mixtures have shown that some principal mechanisms in foam film stability have not been understood so far. For example, the foam film stability above the IEP is independent of the surface coverage and the elasticity. Yet, many studies stress the impact of surface elasticity on the foam film stability. However, a general theory is missing concerning the influence of this parameter on the foam film properties. Furthermore, the elasticity of the foam film interface is not accessible with the present techniques, so that in this work, the difference between the surface characteristic of the bulk solution and that of the film interface was neglected. Nevertheless, theoretical calculations postulate changes of the surface coverage, when a film is formed. To solve these questions, the development of new techniques is required that makes is possible to measure the foam film elasticity directly.

Another open question besides the surface elasticity is the composition of the surface layer. In contrast to insoluble monolayers that are spread on the surface, the surface composition of water-soluble mixtures can not be determined directly from the foam film measurements. To gain more information about the interfacial composition, the combination of TFPB with other techniques like, for example, sum frequency generation would be needed. This would be very helpful for example, in the investigation of catanionic surfactant system to see the transition between the effect of a cosurfactant/organic salt and a surfactant.

References

1. Langevin, D. *ChemPhysChem* **2008**, *9*, 510–522.
2. Exerova, D.; Kruglykov, P. M. *Foam and Foam Films; Theory, Experiment, Application*; Elsevier, Amsterdam, 1998.
3. Bergeron, V. *J. Phys. Condens. Matter* **1999**, *11*, R215.
4. Langevin, D. *Adv. Colloid Interface Sci.* **2001**, *89-90*, 467–484.
5. Evans, D.; Wennerström, H. *The Colloidal Domain - Where Physicls, Chemistry, Biology and Technology Meet*; Wiley-VCH, 2nd Edition, 1999.
6. Hunter, R. *Foundations of Colloid Science*; Oxford University Press, Oxford, 2nd edition, 2001.
7. Israelachvili, *Intermolecular and Surface Forces*; Academic Press, New York, 2nd edition, 1992.
8. Ciunel, K.; Armelin, M.; Findenegg, G. H.; von Klitzing, R. *Langmuir* **2005**, *21*, 4790–4793.
9. Hänni-Ciunel, K.; Schelero, N.; von Klitzing, R. *Faraday Discussion* **2009**, *141*, 41–53.
10. Derjaguin, B.; Chruaev, N.; Muller, V. *Surface Forces*; Consultants Bureau, New York, 1887.
11. Israelachvili, J.; Wennerström, H. *J. Phys. Chem.* **1992**, *96*, 520–513.
12. Gutsche, C.; Keyser, U.; Kegler, K.; Kremer, F.; Linse, P. *Phys. Rev. E* **2007**, *76*, 031403.
13. Butt, H.-J.; Graf, K.; Kappl, M. *Physics and Chemistry of Interfaces*; Wiley-VCH; Weinheim, 2nd edition, 2003.
14. Israelachvili, J. *Intermolecular and Surface Forces*; Academic Press, San Diego, 1998, 2nd edition.
15. Kolaric, B.; Jaeger, W.; Hedicke, G.; von Klitzing, R. *J. Phys. Chem. B* **2003**, *107*, 8152–8157.
16. Schulze-Schlarmann, J.; Buchavzov, N.; Stubenrauch, C. *Soft Matter* **2006**, *2*, 584–594.
17. Bergeron, V. *Langmuir* **1997**, *13*, 3474–3482.
18. Beattie, J.; Djerdjev, A. M.; Warr, G. *Faraday Discussion* **2009**, *141*, 31–39.
19. Buchavzov, N.; Stubenrauch, C. *Langmuir* **2007**, *23*, 5315–5323.
20. Karraker, K.; Radke, C. *Adv. Colloid Interface Sci.* **2002**, *96*, 231–264.
21. Ritacco, H.; Kurlat, D.; Langevin, D. *J. Phys. Chem. B* **2003**, *107*, 9146–9158.
22. Stubenrauch, C.; von Klitzing, R. *J. Phys. Condens. Matter* **2003**, *15*, R1197–R1232.
23. Bergeron, V.; Claesson, P. *Adv. Colloid Interface Sci.* **2002**, *96*, 1–20.
24. Taylor, D. J. F.; Thomas, R. K.; Penfold, J. *Langmuir* **2002**, *18*, 4748–4757.
25. Asnacios, A.; von Klitzing, R.; Langevin, D. *Colloids Surf. A* **2000**, *167*, 189–197.
26. Asnacios, A.; Langevin, D.; Argillier, J. F. *Eur. Phys. J. B* **1998**, *5*, 905–911.
27. Asnacios, A.; Langevin, D.; Argillier, J. F. *Macromolecules* **1996**, *29*, 7412–7417.

28. Stubenrauch, C.; Albouy, P. A.; von Klitzing, R.; Langevin, D. *Langmuir* **2000**, *16*, 3206–3213.
29. Bhattacharyya, A.; Monroy, F.; Langevin, D.; Argillier, J. F. *Langmuir* **2000**, *16*, 8727–8732.
30. Lee, Y.; Dudek, A.; Ke, T.; Hsiao, F.; Chang, C. *Macromolecules* **2008**, *41*, 5845–5853.
31. Noskov, B. A.; Loglio, G.; Miller, R. *J. Phys. Chem. B* **2004**, *108*, 18615–18622.
32. Guillot, S.; Delsanti, M.; Desert, S.; D., L. *Langmuir* **2003**, *19*, 230–237.
33. Monteux, C.; Llauro, M.-F.; Baigl, D.; Williams, C. E.; Anthony, O.; Bergeron, V. *Langmuir* **2004**, *20*, 5358–5366.
34. Jain, N. J.; Trabelsi, S.; Guillot, S.; Langevin, D. *Langmuir* **2004**, *20*, 8496–8503.
35. Trabelsi, S.; Langevin, D. *Langmuir* **2007**, *23*, 1248–1252.
36. Monteux, C.; Williams, C. E.; Bergeron, V. *Langmuir* **2004**, *20*, 5367–5374.
37. Taylor, D. J. F.; Thomas, R. K.; Li, P. X.; Penfold, J. *Langmuir* **2003**, *19*, 3712–3719.
38. Zang, J.; Thomas, R. K.; Penfold, J. *Soft Matter* **2005**, *1*, 310–318.
39. Monteux, C.; Williams, C. E.; Meunier, J.; Anthony, O.; Bergeron, V. *Langmuir* **2004**, *20*, 57–63.
40. Monteux, C.; Fuller, G. G.; Bergeron, V. *J. Phys. Chem. B* **2004**, *108*, 16473–16482.
41. Regismond, S.; Winnik, F.; Goddard, E. *Colloids Surf. A* **1996**, *119*, 221–228.
42. Regismond, S.; Gracie, K.; Winnik, F.; Goddard, E. *Langmuir* **1997**, *13*, 5558–5562.
43. Pezennec, S.; Gauthier, F.; Alonso, C.; Graner, F.; Croguennec, T.; Brule, G.; Renault, A. *Food Hydrocolloids* **2000**, *14*, 463–472.
44. Ropers, M.; Novales, B.; Boue, F.; Axelos, M. *Langmuir* **2008**, *24*, 12849–12857.
45. Bergeron, V.; Langevin, D.; Asnacios, A. *Langmuir* **1996**, *12*, 1550–1556.
46. Asnacios, A.; Espert, A.; Colin, A.; Langevin, D. *Phys. Rev. Lett.* **1997**, *78*, 4974–4977.
47. von Klitzing, R.; Espert, A.; Asnacios, A.; Hellweg, T.; Colin, A.; Langevin, D. *Colloids Surf. A* **1999**, *149*, 131–140.
48. von Klitzing, R.; Espert, A.; Colin, A.; Langevin, D. *Colloids Surf. A* **2001**, *176*, 109–116.
49. Toca-Herrera, J. L.; von Klitzing, R. *Macromolecules* **2002**, *35*, 2861–2864.
50. von Klitzing, R.; Kolaric, B. *Progr. Colloid Polym. Sci.* **2003**, *122*, 122–129.
51. Theodoly, O.; Tan, J. S.; Ober, R.; Williams, C. E.; Bergeron, V. *Langmuir* **2001**, *17*, 4910–4918.
52. Qu, D.; Pedersen, J. S.; Garnier, S.; Laschewsky, A.; Möhwald, H.; von Klitzing, R. *Macromolecules* **2006**, *39*, 7364–7371.
53. Kolaric, B.; Jaeger, W.; von Klitzing, R. *J. Phys. Chem. B* **2000**, *104*, 5096–5101.
54. von Klitzing, R.; Kolaric, B.; Jaeger, W.; Brandt, A. *Phys. Chem. Chem. Phys.* **2002**, *4*, 1907–1914.
55. Milling, A.; Kendall, K. *Langmuir* **2000**, *16*, 5106–5115.
56. Kleinschmidt, F.; Stubenrauch, C.; Deacotte, J.; von Klitzing, R.; Langevin, D. *J. Phys. Chem. B* **2009**, *113*, 3972–3980.
57. Rapoport, D. H.; Anghel, D. F.; Hedicke, G.; Möhwald, H.; von Klitzing, R. *J. Phys. Chem. C* **2007**, *111*, 5726–5734.
58. Anghel, D. F.; Toca-Herrera, J. L.; Winnik, F. M.; Rettig, W.; von Klitzing, R. *Langmuir* **2002**, *18*, 5600–5606.

59. von Klitzing, R.; Espert, A.; Colin, A.; Langevin, D. *Structure of Foam Films containing additionally polyelectrolytes in "Foams, Emulsions and Cellular Materials"*; Kluwer Verlag, 1999; pp 73–82.
60. Letocart, P.; Radoev, B.; Schulze, H.; Tsekov, R. *Colloids Surf. A* **1999**, *151*.
61. Qu, D.; Baigl, D.; Williams, C.; Möhwald, H.; Fery, A. *Macromolecules* **2003**, *36*, 6878.
62. Beltran, C.; Guillot, S.; Langevin, D. *Macromolecules* **2003**, *36*, 8506–8512.
63. Beltran, C.; Langevin, D. *Phys. Rev. Lett.* **2005**, *94*, 217803.
64. Heinig, P.; Beltran, C.; Langevin, D. *Phys. Rev. E* **2006**, *73*, 051607.
65. Delorme, N.; Dubois, M.; Garnier, S.; Laschewsky, A.; Weinkamer, R.; Zemp, T.; Fery, A. *J. Phys. Chem. B* **2006**, *110*, 1752–1758.
66. Mysels, K. J.; Jones, M. N. *Discuss Faraday Soc.* **1966**, *42*, 42–50.
67. Scheludko, A.; Exerowa, D. *Commun. Dept. Chem. Bulg. Acad. Sci.* **1959**, *7*, 105–113?
68. Scheludko, A.; Exerowa, D. *Kolloid-Zeitschrift* **1960**, *168*, 24–8.
69. Exerowa, D.; Scheludko, A. *Chim. Phys. Bulgarien* **1971**, *24*, 47.
70. Exerowa, D.; Nikolov, A.; Zacharieva, M. *J. Colloid and Interface Science* **1981**, *81*, 419.
71. Scheludko, A. *Adv. Colloid Surf. Sci.* **1967**, *1*, 391–464.
72. Scheludko, A.; Platikanov, D. *Kolloid-Z.* **1961**, *175*, 150.
73. Krustev, R.; Müller, H. *Langmuir* **1999**, *15*, 2134–2141.
74. Dörfler, H.-D. *Grenzflächen und kolloid-disperse Systeme*; Springer-Verlag, Berlin, 2002.
75. Atkins, P.; de Paula, J. *Atkins' Physical Chemistry*; Oxford University Press, Oxford, 7th edition, 2002.
76. Lu, J.; Lee, E.; Thomas, R.; Penfold, J.; Flitsch, S. *Langmuir* **1993**, *9*, 1352–1360.
77. Loglio, G.; Pandolfini, R.; Miller, R.; Makievski, A. V.; Ravera, F.; Liggieri, L. *Novel Methods to Study Interfacial Layers*; Elsevier, Amsterdam, 2001.
78. Rotenberg, Y.; Boruvka, L.; Neumann, A. *J. Colloid Interface Sci.* **1983**, *93*, 169–183.
79. Chen, P.; Kwok, D.; Prokop, R. M.; del Rio, O. I.; Susnar, S. S.; Neumann, A. W. *Studies In Interface Science Series*; Elsevier, Amsterdam, Vol. 6, 2001.
80. Ravera, F.; Ferrari, M.; Liggieri, L. *Colloids Surf. A* **2006**, *282-283*, 210–216.
81. Santini, E.; Ravera, F.; Ferrari, M.; Stubenrauch, C.; Makievski, A.; Krägel, J. *Colloids Surf. A* **2007**, *298*, 12–21.
82. Wüstneck, R.; Moser, B.; Muschiolik, G. *Colloids Surf. B* **1999**, *15*, 263–273.
83. Taylor, D.; Thomas, R.; Hines, J.; Humphreys, K. *Langmuir* **2002**, *18*, 9783–9791.
84. Penfold, J.; Tucker, I.; Thomas, R. K.; Taylor, D. J. F.; Zhang, J.; Zhang, X. L. *Langmuir* **2007**, *23*, 3690–3698.
85. Jain, N. J.; Albouy, P.-A.; Langevin, D. *Langmuir* **2003**, *19*, 8371–8379.
86. Noskov, B. A.; Loglio, G.; Lin, S.-Y.; Miller, R. *J. Colloid Interface Sci.* **2006**, *301*, 386–394.
87. von Klitzing, R. *Adv. Colloid Interface Sci.* **2005**, *114*, 253–266.
88. von Klitzing, R.; Kolaric, B. *Tenside Surfactants Detergents* **2002**, *39*, 247–253.
89. Kolarov, T.; Yankov, R.; Esipova, N. E.; Exerowa, D.; Zorin, Z. M. *Colloid Polym. Sci.* **1993**, *271*, 519–520.
90. Kristen, N.; Simulescu, V.; Vüllings, A.; Laschewsky, A.; Miller, R.; von Klitzing, R. *J. Phys. Chem. B* **2009**, *133*, 7986–7990.

91. Li, B.; Geeraerts, G.; Joos, P. *Colloids Surf. A* **1994**, *88*, 251–266.
92. Kristen, N.; von Klitzing, R. *Soft Matter* **2010**, *6*, 849–861.
93. Chakraborty, T.; Chakraborty, I.; Ghosh, S. *Langmuir* **2006**, *22*, 9905–9913.
94. Lynch, I.; Piculell, L. *J. Phys. Chem. B* **2006**, *110*, 864–870.
95. Mya, K.; Jamieson, A.; Sirivat, A. *Langmuir* **2000**, *16*, 6131–6135.
96. O'Driscoll, B.; Fernandez-Martin, C.; Wilso, R.; Knott, J.; Roser, S.; Edler, K. *Langmuir* **2007**, *23*, 4589–4598.
97. Mishra, N.; Muruganathan, R.; Müller, H.-J.; Krustev, R. *Colloids Surf. A* **2005**, *256*, 77–83.
98. Theodoly, O.; Ober, R.; Williams, C. *Eur. Phys. J. E* **2001**, *5*, 51–58.
99. Alexandrova, L. *Adv Colloid Interf Sci* **2007**, *132*, 33–44.
100. Exerowa, D.; Kolarov, T.; Esipova, N.; Yankov, R.; Zorin, Z. *Colloid Journal* **2001**, *63*, 50–56.
101. Warszynski, P.; Barzyk, W.; Lunkenheimer, K.; Frubner, H. *J. Phys. Chem. B* **1998**, *102*, 10948–10957.
102. Teppner, R.; Haage, K.; Wantke, D.; Motschmann, H. *J. Phys. Chem. B* **2000**, *104*, 11489–11496.
103. Balomenou, I.; Bokias, G. *Langmuir* **2005**, *21*, 9038–9043.
104. Üzüm, C. unpublished results.
105. Fundin, J.; Brown, W.; Iliopoulos, I.; Claesson, P. *Colloid Polym Sci* **1999**, *277*, 25–33.
106. Kotsmar, C.; Arabadzhieva, D.; Khristov, K.; Mileva, E.; Grigoriev, D.; Miller, R.; Exerowa, D. *Food Hydrocolloids* **2008**, *23*, 1169–1176.
107. Kristen, N.; Vüllings, A.; Laschewsky, A.; Miller, R.; von Klitzing, R. *Langmuir* **2010**, DOI:10.1021/la1002463.
108. Panchal, K.; Desai, A.; Nagar, T. *J. Dispersion Sci. Technology* **2006**, *27*, 33–38.
109. Bergeron, V.; Jimenez-Laguna, A.; Radke, C. *Langmuir* **1992**, *8*, 3027–3032.
110. Wang, L.; Yoon, R.-H. *Langmuir* **2004**, *20*, 11457–11464.
111. Espert, A.; von Klitzing, R.; Poulin, P.; Colin, A.; Zana, R.; Langevin, D. *Langmuir* **1998**, *14*, 4251–4260.
112. Vollhardt, D.; Emrich, G. *Colloids Surf. A* **2000**, *161*, 173–182.
113. Vollhardt, D.; Brezesinski, G.; Siegel, S.; Emrich, G. *J. Phys. Chem. B* **2001**, *105*, 12061–12067.
114. Stubenrauch, C.; Khristov, K. *J. Colloid Interface Sci.* **2005**, *286*, 710–718.
115. Li, Z. X.; Bain, C. D.; ; Thomas, R. K.; Duffy, D. C.; Penfold, J. *J. Phys. Chem. B* **1998**, *102*, 9473–9480.
116. Soltero, J.; Puig, J.; Manero, O.; Schulz, P. *Langmuir* **1995**, *11*, 3337–3346.
117. Blute, I.; Jansson, M.; Oh, S.; Shah, D. *J. Amercan Oil Chemist Soc.* **1994**, *71*, 41–46.
118. Exerowa, D.; Kolarov, T.; Khristov, K. H. R. *Colloids Surf.* **1987**, *22*, 161–169.
119. Lunkenheimer, K.; Pergande, H.-J.; Krüger, H. *Rev. Sci. Instrum.* **1987**, *58*, 2313–2316.
120. Real Oliveira, M.; Ferreira, J.; Nascimeno, S.; H.D., B.; M.G., M. *J. Chem. Soc. Faraday Trans.* **1995**, *91*, 3913–3917.
121. Wang, Y.; Marques, E.; Pereira, C. *Thin Solid Films* **2008**, *516*, 7458–7466.
122. Patist, A.; Chhabra, V.; Pagidipati, R.; Shah, R.; Shah, D. *Langmuir* **1997**, *13*, 432–434.

123. Nikolov, A.; Wasan, D. *J. Colloid and Interface Science* **1989**, *133*, 1.
124. Bergeron, V.; Radke, C. *Langmuir* **1992**, *8*, 3020.
125. Vassilieff, C.; Nickolova, B.; Manev, E. *Colloid Polym Sci* **2008**, *286*, 475–480.
126. Holmberg, K.; Jönsson, B.; Kronberg, B.; Lindmann, B. *Surfactants and Polymers in Aqueous Solution*; John Wiley & Sons, England, 2nd edition, 2002.
127. Ertekin, A.; Kim, Y.; Kausch, C.; Thomas, R. *J Colloid Interf Sci* **2009**, *336*, 40–45.
128. Knoben, W.; Besseling, N.; Stuart, M. C. *Physicla Review Letters* **2006**, *97*, 068301.
129. Knoben, W.; Besseling, N.; Stuart, M. C. *Langmuir* **2007**, *23*, 6095–6105.
130. Biggs, S.; Burns, J.; Yan, Y.; Jameson, G.; Jenkins, P. *Langmuir* **2000**, *16*, 9242–9248.
131. von Klitzing, R.; Kolaric, B. *Tenside Surfactants Detergents* **2002**, *39*, 247–253.
132. Jönsson, B.; Broukhno, A.; Forsman, J.; Akesson, T. *Langmuir* **2003**, *19*, 9914–9922.
133. Noskov, B. A. *Curr. Op. Colloid Interface Sci.* **2010**, DOI:10.1016/j.cocis.2010.01.006.
134. Milling, A. *J. Phys. Chem.* **1996**, *100*, 8986.
135. Klapp, S.; Zeng, Y.; Qu, D.; von Klitzing, R. *Phys. Rev. Lett.* **2008**, *100*, 118303.
136. Strychalski, E.; Levy, S.; Craighead, H. *Macromolecules* **2008**, *41*, 7716–7721.
137. Pennathur, S.; Baldessari, F.; Santiago, J. G.; Kattah, M. G.; Steinman, J. B.; Utz, P. J. *Anal. Chem.* **2007**, *79*, 8316–8322.
138. Pristinski, D.; Kozlovskaya, V.; Sukhishvili, S. *J. Chem. Phys.* **2005**, *122*, 014907.
139. Cong, R.; Temyanko, E.; Russo, P.; Edwin, N.; Uppu, R. *Macromolecules* **2006**, *39*, 731–739.
140. Krichevsky, O.; Bonnet, G. *Rep. Prog. Phys.* **2002**, *65*, 251–297.
141. Hess, S.; Webb, W. *Biophys. J.* **2002**, *83*, 2300–2317.
142. Jachimska, B.; Jasinski, T.; Warszynski, P.; Adamczyk, Z. *Colloids Surf. A* **2010**, *355*, 7–15.
143. Ferse, B.; Richter, S.; Eckert, F.; Kulkarni, A.; Papadakis, C.; Arndt, K.-F. *Langmuir* **2008**, *24*, 12627–12635.
144. Serghei, A.; Kremer, F. *Rev. Sci. Instrum.* **2006**, *77*, 116108.
145. Israelachvili, J.; Adams, G. *J. Chem. Soc., Faraday Trans.* **1978**, *74*, 975–1001.
146. Tonck, A.; Georges, J. M.; Loubet, J. L. *J. Colloid Interface Sci.* **1988**, *126*, 150–163.
147. Schob, A.; Cichos, F. *J. Phys. Chem. B* **2006**, *110*, 4354–4358.
148. Montiel, D.; Cang, H.; Yang, H. *J. Phys. Chem.* **2006**, *110*, 19763–19770.
149. Savin, T.; Doyle, P. *Biophys. J.* **2005**, *88*, 623–638.
150. Metwalli, E.; Haines, D.; Becker, O.; Conzone, S.; Pantano, C. *J. Colloid Interface Sci.* **2006**, *298*, 825–831.
151. Han, Y.; Alsayed, A.; Nobili, M.; Yodh, A. *Phys. Rew. E* **2009**, *80*, 011403.
152. Froltsov, V.; Klapp, S. *J. Chem. Phys.* **2006**, *124*, 134701.

Acknowledgement

The presented thesis was elaborated during my time as a PhD student at the Stranski Laboratory for Physical and Theoretical Chemistry at the TU Berlin.

I would like to thank my supervisor Prof. Regine von Klitzing for her constant and friendly support and for many helpful discussions and suggestions. I am very grateful that she gave me the opportunity to work in this institute and to present my work at various conferences. She gave me a lot of freedom concerning my scientific work and created a kind atmosphere in the group. I really enjoyed working in this group for the last years.

A big thank you goes to Dr. Reinhard Miller from the Max Planck Institute for Colloids and Interfaces, who adopted me during the time I spent at the MPI to do the surface elasticity measurements and who was always there to help me with the interpretation of the results. I would also like to thank Dr. Csaba Kotsmar, who introduced me to the PAT1 and helped me to solve various technical problems.

Prof. Christine Papadakis from the TU München is acknowledged for giving me the opportunity to conduct FCS measurements in her group and for useful advice concerning the results.

I also thank Prof. Frank Chichos and Martin Pumpa from the University Leipzig for the enormous support with the SFA measurements. They always made my stays in Leipzig very pleasant even beyond science.

I would like to acknowledge Dr. René Strassnick, Rolf Kunert, and Detlef Klabunde for servicing the Thin Film Pressure Balance and for always responding to my emergency calls. Gaby Hedicke is acknowledged for the kind support with the lab work and Andrea Vüllings and Johannes Hellwig for the help with the surface tension measurements.

Many thanks go to my colleges for the friendly and kind atmosphere in the lab. Special thanks go to 'the girls' Natascha, Nora and Anna for encouraging me through the years and for reminding me that there is more in life than science. I also would like to mention Matthias, who gave me a lot of support even though he is on the other side of the planet.

Last but not least, I would like to thank my parents and my husband Paul for constantly supporting and encouraging me on my way.

The financial support from the Deutsche Forschungsgemeinschaft within the framework of the priority program Sfb 448 is kindly acknowledged.

I want morebooks!

Buy your books fast and straightforward online - at one of world's fastest growing online book stores! Environmentally sound due to Print-on-Demand technologies.

Buy your books online at
www.morebooks.shop

Kaufen Sie Ihre Bücher schnell und unkompliziert online – auf einer der am schnellsten wachsenden Buchhandelsplattformen weltweit! Dank Print-On-Demand umwelt- und ressourcenschonend produziert.

Bücher schneller online kaufen
www.morebooks.shop

KS OmniScriptum Publishing
Brivibas gatve 197
LV-1039 Riga, Latvia
Telefax: +371 686 204 55

info@omniscriptum.com
www.omniscriptum.com

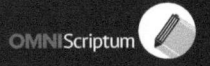

Printed by Books on Demand GmbH, Norderstedt / Germany